Biodiversity and Evolution

Series Editor
Marie-Christine Maurel

Biodiversity and Evolution

Edited by

Philippe Grandcolas
Marie-Christine Maurel

ELSEVIER

First published 2018 in Great Britain and the United States by ISTE Press Ltd and Elsevier Ltd

ISTE Press Ltd
27-37 St George's Road
London SW19 4EU
UK

www.iste.co.uk

Elsevier Ltd
The Boulevard, Langford Lane
Kidlington, Oxford, OX5 1GB
UK

www.elsevier.com

Notices

Knowledge and best practice in this field are constantly changing. As new research and experience broaden our understanding, changes in research methods, professional practices, or medical treatment may become necessary.

Practitioners and researchers must always rely on their own experience and knowledge in evaluating and using any information, methods, compounds, or experiments described herein. In using such information or methods they should be mindful of their own safety and the safety of others, including parties for whom they have a professional responsibility.

To the fullest extent of the law, neither the Publisher nor the authors, contributors, or editors, assume any liability for any injury and/or damage to persons or property as a matter of products liability, negligence or otherwise, or from any use or operation of any methods, products, instructions, or ideas contained in the material herein.

For information on all our publications visit our website at http://store.elsevier.com/

British Library Cataloguing-in-Publication Data
A CIP record for this book is available from the British Library
Library of Congress Cataloging in Publication Data
A catalog record for this book is available from the Library of Congress
ISBN 978-1-78548-277-9

Contents

Introduction xi

Philippe GRANDCOLAS

Chapter 1. From Richard Owen to Charles Darwin:
Understanding the Origin of Life's Diversity. 1

Claudine COHEN

1.1. Introduction 1
    1.1.1. The legacy of German *Naturphilosophie* 3
    1.1.2. French debates over the unity and diversity
    of living beings. 5
1.2. Richard Owen, the archetype and the diversity
of vertebrates 6
    1.2.1. From archetypes to ancestors: Darwin as
    a reader of Richard Owen 8
1.3. Conclusion 10
1.4. Bibliography 11

Chapter 2. Life Engineering in an Evolutionary World. 15

Thomas HEAMS

2.1. Why "Engineers"? 15
2.2. The animal–machine legacy 16
2.3. Genetic engineering: rational design or tinkering? 18
2.4. Synthetic biology as the paradigm of bioengineering. 19
2.5. Transforming a transformation? 21
2.6. Our cousins, the engineers 23
2.7. Engineering and evolutionary dynamics. 24
2.8. Bibliography 26

Chapter 3. The View of Systematics on Biodiversity   29

Philippe GRANDCOLAS

3.1. Introduction   29
3.2. Species: all different.   30
3.3. How about studying the other 90%?   32
3.4. Biodiversity changes   34
3.5. Challenging decades.   35
3.6. Bibliography   36

Chapter 4. Which Model(s) Explain Biodiversity?   39

Guillaume ACHAZ

4.1. Birth–death processes   41
4.2. Coalescent trees   46
4.3. Birth–death and/or coalescent model?   54
4.4. Acknowledgements   56
4.5. Bibliography   56

**Chapter 5. Analysis of Microbial Diversity: Regarding the
(Paradoxical) Difficulty of Seeing Big in Metagenomics**   63

Chloé VIGLIOTTI, Philippe LOPEZ and Eric BAPTESTE

5.1. Introduction   63
5.2. Comparing metagenomic data sets is difficult.   65
5.3. Path dependency and knowledge production   68
5.4. Standardizing metagenomics   73
5.5. Unlocking metagenomics   75
5.6. Conclusion   78
5.7. Acknowledgements   79
5.8. Figure legends   80
   5.8.1. Figure 5.1. Impact of the sequencing method
   on the taxonomic composition of lizard microbiota.   80
   5.8.2. Figure 5.2. Hypothetical sequence of the stages making
   it possible to acquire knowledge in a scientific
   discipline, adapted from Sydow *et al.*   80
   5.8.3. Figure 5.3. Connected components from a similarity
   network between reads of a lizard intestinal microbiome   81
5.9. Bibliography   81

Chapter 6. Genetic Code Degeneracy and Amino Acid Frequency in Proteomes

Chapter 6. Genetic Code Degeneracy and Amino
Acid Frequency in Proteomes.   89
Jean LEHMANN

6.1. Introduction   89
6.2. Frequency–mass correlation of encoded amino acids   92
6.3. Amino-acid volume correlation in the genetic code   93
6.4. Origin of genetic code degeneracy.   96
6.5. Origin of the frequency–mass correlation   100
6.6. Summary and discussion   102
6.7. Conclusion   104
6.8. Acknowledgments.   104
6.9. Bibliography   105

Chapter 7. Telomeres and Telomerases:
Structural Diversity for the Same Role   109
Carole SAINTOMÉ

7.1. Introduction   109
7.2. Nature of chromosome extremities.   111
7.3. Telomeres in eukaryotes   112
  7.3.1. Structures of sequences rich in G: G-quadruplexes   112
  7.3.2. t-loop.   114
  7.3.3. Proteins at telomeres   114
7.4. Maintenance of telomeres by telomerase in eukaryotes   116
  7.4.1. Structure of the catalytic subunit: very
  conserved domain of the RT   117
  7.4.2. RT stretching mechanism   117
  7.4.3. RNA subunit structure   118
7.5. Bibliography   121

Chapter 8. Globalization and Infectious Diseases   123
Thierry WIRTH

8.1. Introduction   123
8.2. Origins of tuberculosis   125
8.3. Emergence of MDR-Beijing lineage.   126
8.4. Staphylococcus aureus and globalization   130
8.5. Intensive farming and its distorting practices   130
8.6. Success of an American clone   132
8.7. Typhoid fever   132
8.8. Indian diaspora and emergence of a new pathogen.   133
8.9. Prospects.   135
8.10. Bibliography   135

Chapter 9. Why are *Morpho* Blue?   139
Vincent DEBAT, Serge BERTHIER, Patrick BLANDIN, Nicolas CHAZOT,
Marianne ELIAS, Doris GOMEZ and Violaine LLAURENS

   9.1. Introduction   139
   9.2. Structural explanation: the iridescent blue
   in *Morpho* is a physical color   142
      9.2.1. Scale structure.   147
      9.2.2. Scale development   150
   9.3. Historical explanation: evolutionary origin
   of blue color in *Morpho*   152
      9.3.1. Color variation in the genus *Morpho*   152
   9.4. Functional explanation: the role of selection in
   the evolution of *Morpho* color.   156
      9.4.1. Thermoregulation.   156
      9.4.2. Hydrophobia.   157
      9.4.3. Signaling to predators: a confusing effect?   157
      9.4.4. Signaling to predators: an aposematic blue?   159
      9.4.5. Sexual selection   160
      9.4.6. Different natural selection between sexes?   162
      9.4.7. And the lack of blue?   162
   9.5. Conclusions and open questions   164
   9.6. Acknowledgments   164
   9.7. Bibliography   165

**Chapter 10. Biodiversity in Natural History Collections:
a Source of Data for the Study of Evolution**   175
Romain NATTIER

   10.1. Introduction.   175
   10.2. Description of biodiversity   177
      10.2.1. Identification, comparison with specimen types   177
      10.2.2. Phylogenetic position   177
      10.2.3. Delimitation of species   178
   10.3. Ecological and evolutionary processes on a population level   178
   10.4. Ecological and evolutionary processes on a phylogenetic scale   180
   10.5. Conclusion   181
   10.6. Acknowledgements   182
   10.7. Bibliography   182

**Chapter 11. Mice and Men: an Evolutionary
History of Lassa Fever** 189
Aude LALIS and Thierry WIRTH

11.1. Symptoms and methods of prevention 190
11.2. Towards a better understanding of the viral reservoir:
phylogenetic systematic gene flow and phenotypic variation. 193
11.3. Lassa virus, molecular evolution and dating 203
11.4. Role of armed conflicts and refugee movements
in the spread of Lassa fever in West Africa 204
11.5. Viral epidemiology: towards an integrated approach 205
11.6. Bibliography 206

**Chapter 12. Evolutionary History of Moles in Western
Europe: One Mole May Hide Another!** 213
Violaine NICOLAS, Jessica MARTINEZ-VARGAS and Jean-Pierre HUGOT

12.1. An unexpected biodiversity 213
12.2. A new mole species in France and Spain? 215
12.2.1. Mitochondrial DNA: three lineages of *T. europaea* 216
12.2.2. Nuclear DNA: *T. europaea* is paraphyletic 217
12.2.3. Morphological analysis: a new mole species 219
12.3. Factors affecting the geographical distribution
of the species in France. 220
12.4. Bibliography 223

**Chapter 13. The Conoidea and Their Toxins:
Evolution of a Hyper-Diversified Group.** 227
Nicolas PUILLANDRE and Sébastien DUTERTRE

13.1. General introduction and state of the art 227
13.1.1. Systematics. 228
13.1.2. Toxinology. 230
13.2. Recent developments in systematics 234
13.2.1. Species delimitation. 234
13.2.2. Phylogeny and classification 237
13.3. Toxins: genomic, transcriptomic and proteomic approaches 238
13.3.1. Methodological developments 239
13.3.2. Recent discoveries. 241
13.4. Evolution of Conoidea: integrative approaches 242
13.5. Bibliography 244

**Chapter 14. The Anthropocene: a Geological
or Societal Subject?**   251
Patrick DE WEVER and Stan FINNEY

14.1. A new "geological era"?   253
14.2. Criteria to distinguish the Anthropocene.   255
14.3. Why is the issue of the Anthropocene being imposed?   260
14.4. Conclusion   261
14.5. Bibliography   263

List of Authors   265

Index   269

Introduction

Evolution and Biodiversity

The word biodiversity is fashionable. This term, which is a contraction of biological diversity, was penned by Thomas Lovejoy, Walter G. Rosen and Edward Osborne Wilson to refer to the living world without any restriction of aspect or study level [WIL 88]. In the dramatic context of the sixth extinction, the word was so popularized by countless publications that it became well known to all. In only a few decades, the word – biodiversity – then replaced another one – the living world – to identify this essential element of our environment. It focuses on the differences between individuals (variety) rather than their main common points ("living world laws"). After our altogether recent discovery of biological laws (for example, heredity), we have learned since to study their consequences in our environment as a whole, no longer focusing on our laboratory organisms.

Even so and in terms of semantics, the word biodiversity does not integrate all aspects of the living world. One of the main characteristics of biodiversity is that it is alive and, as such, it permanently changes. The variations between individuals in populations and species result from their differentiation during biological evolution. To understand and interpret it, we must reconstruct the paths followed by evolution, by tracking them among kinship relationships (phylogenies) that can be inferred between individuals. To understand a trajectory, must we not draw a line between a point of origin and stages?

Introduction written by Philippe GRANDCOLAS.

Furthermore, although most people understand that the present variety is the result of past evolution, few of us take into account that the organisms we see are constantly changing. In fact, at our timescale, many species seem the same in the main lines of their phenotypes; however, individuals are in fact subject to selection pressure variations, gene transfers, random mutations, development mechanics and, finally, interactions with the environment, including the biological environment. In short, no species is unchanging and we realize it when we delve into the details of their genome or their anatomy at the level of a few generations. By releasing or exercising a selection pressure, we will see a previously stable state, which we thought was fixed, gradually changing.

Biological evolution, which we think results from a long process marked by a series of deceiving ancestral figures, is in fact a permanent phenomenon, characterized by the constant reproduction and transmission of modified ancestral traits to descendant. The best proof of evolution in the eyes of the common people should reside in the composite likeness of their children (how many creationists are thus proud of their offspring, without suspecting their apostasy!).The inheritable appearance of "resistances" over a few generations in bacteria or insects, in other words metabolic or physiologic changes enabling them to deal with antagonist substances (antibiotics, insecticides, etc.), is another striking example.

Consequently, our view of biodiversity must take into account this past and future. "Nothing in biology makes sense except in the light of evolution", to quote Theodosius Dobzhansky [DOB 73], and nothing in the future of biodiversity makes sense if we do not understand that it constantly changes and adjusts in a subtle and often little predictable way.

This book focuses on illustrating this duality of the living world, which reproduces and transmits its ancestral traits to descendant, but changes despite everything. In addition to improvement of our knowledge, studying biodiversity and evolution is a valuable school of ethics. Faced with other organisms, we put into perspective our human states and behaviors. We face evolutionary change, and we learn that things are not immutable. Finally, we face the consequences of our action on the environment and maybe we learn something from all these retroactive effects that affect us directly.

Bibliography

[DOB 73] DOBZHANSKY T., "Nothing in Biology Makes Sense Except in the Light of Evolution", *American Biology Teacher,* vol. 35, pp. 125–129, 1973.

[WIL 88] WILSON E.O. (ed.), *Biodiversity*, National Academies Press, Washington, 1988.

From Richard Owen to Charles Darwin: Understanding the Origin of Life's Diversity

1.1. Introduction

Where does the immense diversity of living beings surrounding us come from? How can we understand its origin? For a long time, creationism was the sole response to these queries. During the Renaissance, collectors and "virtuosi" accumulated in their curiosity cabinets new animals, plants and fruits brought from distant countries by travelers and marveled at their extraordinary diversity, which was generally referred to as God's creative power. By the end of the 17th Century, a whole multitude of unknown beings appeared under the newly invented microscope: these observations, experiments and questions fascinated the public [MOR 11]. "We see with microscopes very small drops of rainwater, vinegar, or other liquors, full of little fish and serpents that we could not ever suspect to inhabit these..." explained Fontenelle to the Marquise of the *Entretiens sur la Pluralité des Mondes* [FON 86, *Troisième soir*]. The observation of infinitely diverse, small and complex living organisms led to theological conclusions: for Swammerdam, the metamorphoses of insects were proofs of God's perfection and of the uniformity of laws in the universe [SWA 82]. Réaumur found in his entomological observations an ever renewed occasion to admire

Chapter written by Claudine COHEN.

divine providence [REA 42]. These observations of nature, of its marvelous perfection and infinite wealth opened, with Abbé Pluche [PLU 32] in France and William Paley [PAL 02] in England, the way to "natural theology", which viewed nature's wonders as the effect of God's design and pervaded natural history well into the 19th Century.

On the other hand, nature revealed an order that could be rationally and systematically described. Beyond traditional Aristotelian taxonomic schemes and classifications inspired by Renaissance "correspondences" [FOU 66], novel classificatory modes attempted to define natural classes of living beings by describing their external features [DAU 26]. During the first decades of the 18th Century, Linnaeus developed his method to build a systematized classification of the living world, first focusing on plants, then on animals and humans [LIN 53, LIN 58]; this new classificatory system could make the diversity of living beings understood, named and organized. Linnaeus' rational classification of the living world also embodied a theological vision of Creation [VEU 08]. This comprehensive view of the living world was associated with the idea of a hierarchic structure of the natural world, systematized by the Leibnizian representation of the "Great Chain of Being" [LOV 34]: for most 18th Century naturalists, this concept of a "series" generally did not reflect, however, a temporal succession of changes but a continuum of resemblances [DAU 26, BAR 88a]. Charles Bonnet's *palingénésie* described the order of the living world as a hierarchical and fixed scale with humans at its top [BON 45]. In the successive volumes of his *Histoire naturelle*, Buffon described the diversity of animals and their adaptations to different "climates" [BUF 88a]: he did admit the possible transformations of living beings, conceived as a "degeneration" under the effect of the environment [BUF 66, 311sq]. Without abandoning a creationist framework, he offered in his last work, *Epochs of Nature* [BUF 88b], a historical vision of nature, and described its successive ages as a series of seven "epochs" culminating in the appearance of Humans, who were created to dominate nature and make it thrive.

By mid-19th Century, Darwin radically broke away from natural theology and the idea of Creation. He proposed revisiting the question of the origin of living beings and their transformations over geological times in materialistic terms. Individual variations, natural selection and adaptation were the main mechanisms at work in this evolutionary process. For Darwin, the extant diversity of the living world, including Mankind, was thus the result of a continuous descent of beings he figured as a tree-like scheme, whose divergent branches were rooted in a common ancestor [DAR 72]. Among the multiple empirical and theoretical sources fueling the "long argument" of Darwin's *Origin of Species*, we will highlight here the importance of morphology and of the notion of *form*, which was developed in natural sciences since the turn of the 19th Century, and led many naturalists to consider life's diversity within a new conceptual framework. Transcendental morphology, which first appeared in Germany and then developed in the whole of Europe, in fact played a crucial part in the emergence of 19th-Century evolutionary thinking. It is this legacy that I will consider here, focusing particularly on the works of British anatomist and paleontologist Richard Owen, although Owen was an opponent to Darwin's evolutionism [DES 82, DES 89, RUP 94]. I will argue that they were in fact essential to the triggering in England of the emergence of a new way of looking at the diversity of living beings and its origin.

1.1.1. The legacy of German *Naturphilosophie*

By the end of the 18th Century, German *Naturphilosophie* laid the foundations for a philosophical, scientific and literary approach to nature and its beings, partly fostered by Kantian philosophy [LEN 81, SLO 92]. At the time of the emergence of comparative anatomy [BAL 79], embryology [GOU 77] and paleontology [LAU 87, COH 11], transcendental morphology relied on the notion of a universal order, the validity of which is not only metaphysical, but also physical. It also implied the idea of a progressive change in nature: the whole nature is in motion, and a trend directed from the lowest to the highest, from chaos to Mankind, is its main character. Man, the greatest being on Earth, is also inextricably connected to all the objects of nature; he embodies the goal towards which everything aims.

Naturphilosophie asserts both and, at the same time, the infinite diversity of nature and its unity, which transcends the vision of the eye and can only be perceived by the mind. It proposes the notion of a morphological model, an ideal matrix accounting for the variability of all living beings [BAL 79, LEN 81, COH 00]. The immense number of living species can be subsumed under one or more ideal types, which provide and reveal the secret logic behind the multiplicity of living forms. From the last decades of the 18th Century onwards, a whole generation of German anatomists, such as Lorenz Oken [OKE 07], Johann Friedrich Meckel [MEC 08], Karl-Friedrich Kielmeyer [KIE 93] and Karl Gustav Carus [CAR 28], but also literary authors, such as Friedrich von Schelling [VON 88] and Johann Wolfgang von Goethe, played a decisive role in forging and spreading these concepts.

Goethe himself tried his hand at anatomical observations and speculations. In his *Versuch über die Gestalt der Thiere* [GOE 90], he made the hypothesis of a "fundamental type" representing the skeletal scheme of all vertebrates. Just as all the organs of a plant can be considered as modifications of the leaf [GOE 90], the different parts of the vertebrate skeleton can be represented as metamorphosed vertebrae. Moreover, Goethe brought to light in the human skull a structure which is common to all vertebrates, the "intermaxillary bone", thus providing evidence of morphological unity between all vertebrates, including humans [GOE 90]. These observations and ideas were widely influential in German natural sciences and were extensively used, for example in Oken's work *On the Meaning of Skull Bones* [OKE 07] and later in the anatomic work of the physician and painter Karl Gustav Carus [CAR 35].

The notion of a single anatomical model informing the whole diversity of living beings was also present in the work of embryologists Kielmeyer and Meckel, who saw in the successive developmental stages of the embryo a summary of the adult forms of inferior animals. In this way, ontogenetic development could provide a key to the hierarchy of living beings. Embryologist Karl Ernst Von Baer later developed a different view, which he exposed from 1828: for him, the ontogeny of higher beings did not recapitulate the sequence of adult, but of embryonic forms [GOU 77]. Ontogenetic development could thus be considered as a gradual process of specialization from a simple structure. Johannes Müller summarized in these words Von Baer's thought: "During the first stages of their formation, vertebrate embryos, in all their purity, present the most general and simplest traits of a vertebrate animal type, and this is what makes them look like each other to such a point that it is often hard to differentiate them from one another. Fish, reptiles, birds, mammals and humans are first the simplest expression of the type common to all; but it gradually strays from it as they develop, and the extremities, for example, after looking the same for some time, turn into fins, wings, hands, feet, etc." [MUL 30:705].

Thus, for the advocates of *Naturphilosophie*, notwithstanding their scientific and philosophical divergences, comparative anatomy and embryology could account for the diversity of beings, and reveal their kinship through a common "plan" from which they were derived. Born in Germany, the notions of a unity of the "composition plan" and of a single biological model informing the structure of all living beings were the source of important developments and debates in Romantic Europe, especially in France and in England.

1.1.2. French debates over the unity and diversity of living beings

In France, zoologist Etienne Geoffroy Saint-Hilaire drew on German *Naturphilosophie* to support the idea that there was a single "pattern" in the living world, a morphological model, from which the whole diversity of living organisms was derived [GEO 22]. At the same time, great French paleontologist George Cuvier, although trained in German comparative anatomy, rejected the principles of transcendental morphology and mocked attempts made by Oken, Carus or Geoffroy, to subsume all beings under a single anatomical model. Cuvier considered living organisms as immutable and finalized closed systems, and he distinguished in the animal kingdom four separate branches (*Vertebra*, *Mollusca*, *Articulata*, *Radiata*) with totally distinct anatomical and functional patterns [CUV 17]. In contrast, Geoffroy insisted on identifying similar anatomic schemes in organisms belonging to different species. his *principle of connection* gave the key to this correspondence. In his view, structures belonging to different organisms could be referred to as the same morphological model, despite changes in size and form, if they bore an identical anatomical relationship with the body's other parts. The coherence of an organism could be understood in reference to a general scheme, to a morphological model: the *Bauplan*, which specified the order through which the different parts of this organism fitted together. This common pattern opened the possibility of a transformation of living organisms: an organism is but one expression of a general scheme, while a multitude of forms matching the same scheme are likely to appear in the course of life's history.

During a famous 1830 debate at the Paris Académie des Sciences over the "unity of composition plan", Cuvier strongly rejected Geoffroy's claim that the body structure of cephalopods was identical to that of vertebrates [APP 87]. If Cuvier then triumphed, it was probably because of his scientific stature and authority, but also his position as a public figure and his talent as a writer, which made him a prominent leader in the French science of his time: his ideas would continue to be cultivated as dogmas until late in the century. However, in the following decades, attempts to reconcile these two opposite visions of the organization of living beings and of life's diversity would occur elsewhere – namely in England.

1.2. Richard Owen, the archetype and the diversity of vertebrates

Richard Owen played, from 1840, the role of a leading scientific authority in English science [RUP 93], a member of the Royal College of Surgeons, a professor of physiology at the Royal Institution, a lecturer at the Ecole des Mines and a superintendent of the Bloomsbury Natural History collections; he created the Natural History Museum of London in 1863. Throughout his life, Owen carried out considerable works in comparative anatomy and paleontology. He studied fossil vertebrates and engaged in an intense activity to reconstruct species. In 1836, Darwin gave him the fossils collected during the Beagle voyage to study [DAR 39]. His *Fossil Vertebrates of England*, published in 1846 in four volumes, was completed by the three volumes of his *Comparative Anatomy and Physiology of Vertebrates* [OWE 68] and by several theoretical works on "anatomical philosophy" [OWE 48].

The legacy of German *Naturphilosophie* persisted in Owen's works. However, although he was a strong supporter of transcendental morphology and Geoffroy's and Carus' friend, Owen also claimed to be Cuvier's successor, perhaps even the "Newton" of paleontology that the *Discourse on the Revolutions of the Surface of the Globe* [CUV 25] had called for. His whole descriptive and theoretical work aimed to reconcile the "two great lights" of French natural history: but how was it possible to reconcile Cuvier's concept of the functional unity of the organism with Geoffroy's notion of a transformation of species and a unique morphological plan of organization for the entire animal kingdom? Owen did not try to argue against Cuvier's division of the animal reign into four branches: instead, he advocated for a single organization plan for each of these main classes. He placed at the center of his theoretical construction and of his own anatomist practice the notion of *homology*, which he defined, according to the principles implemented by Geoffroy, as the variation of the same structure in different organisms, referred to as a single model. In different species, a homologous part occupied the same place in the general organism plan and had the same limits, without necessarily taking on the same function.

Owen's morphological thought was developed in a small incisive essay with a modest title: *On the Nature of the Limbs* [OWE 08]. With an austere and technical form, this book is rich in ideas concerning the structure and diversification of living beings. In it, Owen thoroughly compared the limbs' osteological structure of different classes of vertebrates – fish, amphibians, reptiles, birds, mammals and men – referring them to a common skeletal plan or the same archetype. The concept of "archetype" here used by Owen was a key to account for both the unity and variety of living beings. Owen introduced this notion as early as 1846, in a lecture presented before the British Association for the Advancement of Science [RUP 93]. A single "plan", "model" or "idea" governs the anatomical organization of all vertebrates, he explained. This ideal model, which is the basis of animal organization in one same class, can be recognized through anatomical homologies, such as "the numerous and beautiful evidences for the unity of plan which the structures of the locomotive members have disclosed" [OWE 48].

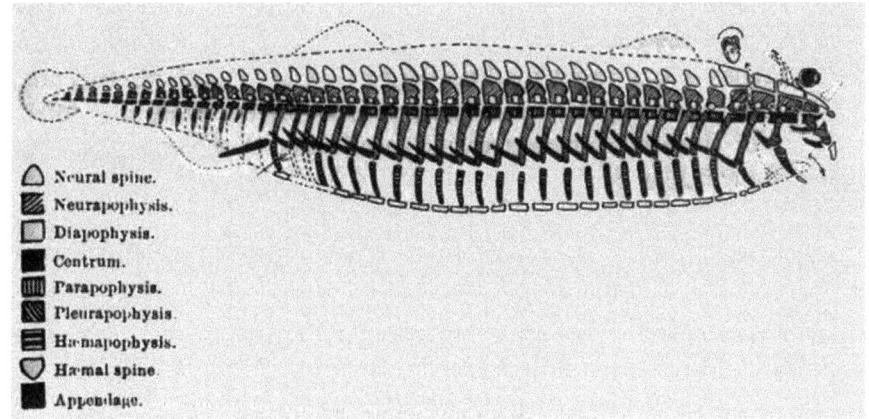

Figure 1.1. *The vertebrate archetype, according to Richard Owen [OWE 08]*

The archetype was, in Owen's own words, a *generalization*: "the highest that can be achieved". The closer an animal was to the archetype, the less it was perfect: the archetype represented the lowest degree of development within a type. To embody the vertebrate archetype, Owen chose a primitive fish, the *Lepidosiren*, whose endo-skeletal structure is composed of a simple succession of unspecialized vertebral segments, indicating the prevalence of what he called the "vegetative force", a principle of repetition. In the organization of every main class of vertebrates, this repetitive force is more or less dominated by an adaptive force, which modified the archetype's elements, thus enabling them to achieve distinctive functions.

Criticism of finality was crucial to Owen's reflection. Morphological structures, he remarked, impose their limits upon the archetype's various expressions: therefore, in different kinds of organisms, the available elements are reused within a set of new adaptations, some anatomical structures being enlarged, reduced or distorted, while others, which are unused, remain as vestigial parts. Therefore, Owen discusses Cuvier's teleological view of the organism as a closed and finalized structure, within which every bone, every organ, has a definite function. This opens the possibility to comprehend both the diversification and transformations of living organisms: for Owen, fauna undergo successive changes. "The creation of each animal class, reptiles, birds and tetrapods, was successive and continuous since the most ancient times" [OWE 58: LXXV]. This successive appearance of species, each of them being a transformation of the archetype, and being well adapted to its environment, could only be understood through the intervention of an "anticipating intelligence", a divine power.

Thus, Owen associated the representation of "nature's harmony" with that of a continuous transformation of species, whose development is referred to as a transcendent and unchanging "type": a view which is both aesthetic and theological. "Derivation sees... a manifestation of creative power in the variety and beauty of the result" [OWE 68: 808], within a living world whose secret cohesion and hierarchy have to be deciphered, and whose laws guarantee both the permanence of beings, their transformations and the creation of an immense diversity of living forms.

1.2.1. *From archetypes to ancestors: Darwin as a reader of Richard Owen*

Owen took part in the heated debates that occurred in mid-century England over Darwin's evolutionary theory. He was familiar with Darwin's work, though he rejected his ideas on evolutionary processes. In his own view of species "derivation", Owen drew neither on Lamarck's habits and will nor on Darwin's contingent mechanism of natural selection. Instead, he believed in an "innate tendency" of beings to metamorphose during the course of the history of species, triggered by God's continuous action on nature and creative power. The anti-Darwinian positions of Owen and other British naturalists of his generation probably reflected in some respects those of the political and social milieu to which they belonged [DES 89]. Their opposition to the notion of a contingent evolution of life, their rejection of natural selection and their effort to maintain Man's dominant place in nature – these positions probably reflected their socio-political background: the defense of the conservative values of the colonial aristocracy and bourgeoisie of the time. However, besides noisy controversies, it should be stressed that considerations on morphology presented in Owen's work played a crucial role in the genesis of Darwin's evolutionary thinking.

In a handwritten note in a margin of his copy of Owen's book [DIG 90] *On the Nature of the Limbs*, Darwin wrote: "I look at Owen's archetypes as more than idea[s/l?], as a real representation as far as the most consummate skill and loftiest generalization can represent the parent form of Vertebrata". In the last chapter of the *Origin of Species* [DAR 72], chap. XIV – "Mutual Affinities of Organic Beings - morphology - embryology - rudimentary organs"), Darwin recapitulates the proofs of evolution derived from morphology – "one of the most interesting departments of natural history, [which] may almost be said to be its very soul", he wrote, as he took up the vocabulary of transcendental morphologists: "We have seen that members of the same class, independently of their habits of life, resemble each other in the *general plan of their organization*. This resemblance is often expressed by the term '*unity of type*'; or by saying that the several parts and organs in the different species of the class are *homologous*" [DAR 72]. If he paid tribute here to German naturalists and to Etienne Geoffroy Saint Hilaire, Darwin also referred (sometimes implicitly) to Owen's brilliant essay [OWE 08]: "What can be more curious than that the hand of a man, formed for grasping, that of a mole for digging, the leg of the horse, the paddle of the porpoise, and the wing of the bat, should all be constructed on the same pattern, and should include similar bones, in the same relative positions? [...] Geoffroy St. Hilaire has strongly insisted on the high importance of relative position or connexion in homologous parts; they may differ to almost any extent in form and size, and yet remain connected together in the same invariable order". Darwin wrote in *The Origin of Species* [DAR 72, Chap. XIV].

Embryology is also cited at length in this same chapter, as bearing evidence to common ancestry. "Embryology rises greatly in interest, when we look at the embryo as a picture, more or less obscured, of the progenitor, either in its adult or larval state, of all the members of the same great class" [DAR 72]. Here, Darwin particularly refers to Von Baer's work: "it has been shown that generally the embryos of the most distinct species belonging to the same class are closely similar, but become, when fully developed, widely dissimilar. A better proof of this latter fact cannot be given than the statement by Von Baer that the embryos of mammalia, of birds, lizards, and snakes, probably also of chelonia, are in their earliest states exceedingly like one another, both as a whole and in the mode of development of their parts; so much so, in fact, that we can often distinguish the embryos only by their size".

In his tribute to morphology, Darwin also included a criticism of the notion of "final causes", which seems directly inspired by Owen. "Nothing can be more hopeless than to attempt to explain this similarity of pattern in members of the same class, by utility or by the doctrine of final causes. The hopelessness of the attempt has been expressly admitted by Owen in his most interesting work on the 'Nature of Limbs'".

1.3. Conclusion

The relationship between Owen and Darwin, although difficult and confrontational – and now somewhat forgotten – was important for the development of concepts making it possible to understand the logical/temporal origin of life's diversity. As we have seen, in Darwin's thought, comparative morphology and embryology offered crucial evidence of the demonstration of evolution [DAR 72, XIV]. "It was easy for Darwin to transfer the volume of Owen's work on comparative anatomy of vertebrates into a remarkable synthesis of evolution through natural selection", Nicolas Rupke wrote. "In Darwin's reinterpretation, homology became proof of ancestry and, conversely, ancestry became the criterion for homology" [RUP 94: 210].

The notion of *origin* could now be understood on several levels: in Darwin's vocabulary, the "*origin* of species" characterizes a process through which species are differentiated through descent with modification. The *origin* of a family or a genus is its "common ancestor", a form that can be revealed through comparative morphology, embryology and paleontological research. Finally, the *origin* of life could be hypothesized, Darwin writes, as "...a small number of forms, or even... only one... a simple beginning" from which the immense diversity of the living world proceeds. In Darwin's thinking, origin no longer refers to a mystical archetype or a theology of Creation: common ancestors are the genealogical and material foundation of beings' diversification. Thus Darwin operates a remarkable reversal of Owen's thinking: replacing Archetypes by ancestors opens a whole new vision of the world, in which the action of a Creator is no longer needed to account for life's diversity and its origin.

1.4. Bibliography

[APP 87] APPEL T., *The Cuvier-Geoffroy Debate*, Oxford University Press, Oxford, 1987.

[BAL 79] BALAN B., *l'Ordre et le temps: l'anatomie comparée et l'histoire des êtres vivants au 19ᵉ siècle*, Vrin, Paris, 1979.

[BAR 88] BARSANTI G., "Les immagini della natura: scale, mappe, alberi, 1700–1800", in *Nuncius, Annali di storia della scienza*, Firenze, Anno III, fasc. 1, pp. 55–125, 1988.

[BON 45] BONNET C., *Traité d'insectologie*, Durand, Paris, 1745.

[BUF 66] BUFFON G.-L., *Histoire naturelle, générale et particulière: De la Dégénération des animaux*, vol. 15, Imprimerie royale, Paris, 1766.

[BUF 88a] BUFFON G.-L., *Histoire naturelle, générale et particulière*: vol. 36, Imprimerie royale, Paris, 1749–1788.

[BUF 88b] BUFFON G.-L., *Histoire naturelle, générale et particulière, Les Epoques de la Nature*, Imprimeric royale, Paris, 1788.

[CAR 28] CARUS K.G., *Grundzüge der vergleichenden Anatomie auf Physiologie*, Hilschersche Buchhandlung, Dresden, 1828.

[CAR 35] CARUS K.G., *Traité élémentaire d'anatomie comparée*, J.B. Baillière, Paris, 1835.

[COH 00] COHEN C., "Richard Owen: Paléontologie, embryologie et morphologie transcendantale autour de 1840" in BLOCH O. (ed), *Les Philosophies de la nature*, Presses de la Sorbonne, Paris, 2000.

[COH 11] COHEN C., *La Méthode de Zadig: La trace, le fossile, la preuve*, Editions du Seuil, Paris, 2011.

[CUV 17] CUVIER G., *Le Règne animal distribué selon son organisation*, Fortin, Masson et Ciè, Paris, 1817.

[CUV 25] CUVIER G., *Discours sur les révolutions de la surface de globe*, Dufour, Paris, 1825.

[DAR 39] DARWIN C., *The zoology of the voyage of H.M.S. Beagle, under the command of Captain Fitzroy, during the years 1832 to 1836*, Part 1, No. 2: Fossil Mammalia, Smith, Elder and Co., London, 1839.

[DAR 72] DARWIN C., *On The Origin of Species*, 6th edition, John Murray, London, 1872.

[DAU 26] DAUDIN H., *De Linné à Jussieu, Méthode de classification et idée de série*, Felix Alcan, Paris, 1926.

[DES 82] DESMOND A., *Archetypes and Ancestor: Palaeontology in Victorian London, 1850–1875*, Chicago University Press, Chicago, 1982.

[DES 89] DESMOND A., *The Politics of Evolution: Morphology, Medicine and Reform in Radical London*, Chicago University Press, Chicago, 1989.

[DIG 90] DI GREGORIO M., GILL N. (eds), *Charles Darwin's Marginalia*, Garland Press, New York, 1990.

[FON 86] FONTENELLE B., *Entretiens sur la Pluralité des Mondes*, E. Didier, Paris, 1686.

[FOU 66] FOUCAULT M., *Les Mots et les Choses*, Gallimard, Paris, 1966.

[GEO 22] GEOFFROY SAINT HILAIRE, *Philosophie anatomique*, J.B. Baillière, Paris, 1822.

[GOE 90] GOETHE J.W., *Urfplanze*, in KUHN D. (ed.), *Sammtliche Werke, vol. 40: Naturkundliche Schriften II, Schriften zur Morphologie*, Deutscher Klassiker Verlag, Frankfurt am Main, 1990.

[GOE 95] GOETHE J.W., *Über den Zwischenkiefer des Menschen und der Tiere* (1784), in *Goethe, Scientific Studies*, Princeton University Press, Princeton, 1995.

[GOU 77] GOULD S.J., *Ontogeny and Phylogeny*, Harvard University Press, Cambridge 1977.

[KIE 93] KIELMEYER C.-F., *Ueber die Verhaltnisse der organischen Krafte, unter einander in der Reihe der verschiedenen Organisationen, die Geseze und Folgen dieser Verhaltnisse : ein Rede den 11ten Februar 1793, am Geburstage des regierenden Herzogs Carl von Wirtemberg im grossen akademischen Horsale gehalten*, Mit Akademischen Schriften, Stuttgart, 1793.

[LAU 87] LAURENT G., *Paléontologie et évolutionnisme de Cuvier-Lamarck à Darwin*, Editions du CTHS, Paris, 1987.

[LEG 97] LE GUYADER H., *Etienne Geoffroy Saint-Hilaire*, Belin, Paris, 1997.

[LEN 81] LENOIR T., "The Göttingen School and the development of Transcendantal Naturphilosophie in the Romantic Era", *Studies in the History of Biology*, vol. 5, pp. 111–205, 1981.

[LIN 53] LINNAEUS K.V., *Species plantarum*, Laurentius Salvius, Stockholm, 1753.

[LIN 58] LINNAEUS K.V., *Systema naturae*, Laurentius Salvius, Stockholm, 1758.

[LOV 34] LOVEJOY O., *The Great Chain of Being*, Harvard University Press, Cambridge, 1934.

[MEC 08] MECKEL J.F., *Beyträge zur vergleichenden Anatomie*, vol. 2, C.H. Reclam, Leipzig, 1808.

[MOR 11] MORNET D., *Les sciences de la nature au XVIIIe siècle*, Armand Colin, Paris, 1911.

[MUL 30] MULLER J., *Bildungsgeschichte* der *Genitalien aus anatomischen Untersuchungen an Embryonen des Menschen und der Thiere*, Arnz, Düsseldorf, 1830.

[OKE 07] OKEN L., *Ueber die Bedeutung der Schädelknochen*, Göbhardt, Bamberg, 1807.

[OSP 76] OSPOVAT D., "The Influence of Karl Ernst's von Baer's Embryology, 1828–1859", *Journal of the History of Biology*, vol. 9, pp. 1–28, 1976.

[OWE 48] OWEN R., *On the Archetype and Homologies of the Vertebrate Skeleton*, Richard and John E. Taylor, London, 1848.

[OWE 58] OWEN R., *Presidential* Address, *Report of the British Association for the Advancement of Science*, Leeds, 1858.

[OWE 68] OWEN R., *Comparative Anatomy and Physiology of Vertebrates*, Longmans, London, 1868.

[OWE 92] OWEN R., *The Hunterian Lectures in comparative anatomy, May and June 1837*, University of Chicago Press, Chicago, 1992.

[OWE 08] OWEN R., *On the Nature of the Limbs: A Discourse*, University of Chicago Press, Chicago, 2008.

[PAL 02] PALEY W., *Natural Theology, Or Evidences of the Existence and Attributes of the Deity*, J. Faulder, London, 1802.

[PLU 32] PLUCHE A., *le Spectacle de la Nature*, Imprimerie des Célestins, Bar-le-Duc, 1732.

[RÉA 42] RÉAUMUR A.F., *Mémoires pour servir à l'histoire des Insectes*, Académie Royale des Sciences, Paris, 1742.

[ROG 68] ROGER, J., *Les Sciences de la Vie dans la pensée française du XVIIIe siècle*, Masson, Paris, 1968.

[RUD 72] RUDWICK M., *The Meaning of Fossils: Essays in the History of Palaeontology*, Elsevier, London, 1972.

[RUD 97] RUDWICK M., *Georges Cuvier, Fossil Bones and Geological Catastrophes*, University of Chicago Press, Chicago, 1997.

[RUP 93] RUPKE N.A., "Richard Owen's Vertebrate Archetype", *ISIS*, vol. 84, pp. 231–251, 1993.

[RUP 94] RUPKE N.A., *Richard Owen, Victorian Naturalist*, Yale University Press, New Haven, 1994.

[SWA 82] SWAMMERDAM J., *Histoire générale des Insectes*, Jean Ribbius, Utrecht, 1682.

[VEU 08] VEUILLE M., DROUIN J-M., DELEPORTE P. *et al.* (eds), "Linnaeus, Systématique et biodiversité", *Biosystema,* vol. 25, 2008.

[VON 88] VON SCHELLING F., *Ideas for a Philosophy of Nature: as Introduction to the Study of this Science*, Cambridge University Press, Cambridge, 1988.

2

Life Engineering in an Evolutionary World

2.1. Why "Engineers"?

The first scene of Ridley Scott's science fiction film Prometheus (2012) presents an alien decomposing in order to sow a planet that could be Earth with DNA and give birth to life. We learn later on in the film that the alien is part of a species that came from space called the Engineers. Why has such a term made its way from technique to fiction via science? This chapter could indeed be an attempt to answer this question. Our conviction would be that the explanation is partly based on the increasing popularity of this word in the last few years and that expressions such as living engineering, genetic engineering, metabolic engineering and even ecosystem engineering are particularly popular, describing different study subjects and involving meanings that are very different from that of the very term, to such a point that it seems that the time has come to have a critical look at its uses.

As a starting point, we will readily agree that a word can be pragmatically used, have a certain metaphorical evocation power and even be entitled to a certain polysemy in different contexts. Therefore, we will not subject the reader here to us being pointlessly offended and rejecting in advance any transposition of the word "engineering" outside its original context. We will even immediately highlight, in the cases we will review, that one of the retrospective arguments in favor of its use is the actual enthusiasm it generates: it is likely that genetic engineering or ecosystem engineering would not have had, as expressions, its undeniable impact on the respective scientific communities, if these expressions *only* seemed paradoxical or bold

Chapter written by Thomas HEAMS.

and if they did not echo the realities perceived by these communities and the needs for formalization arising from it. In fact, rigor and scientificity are not measured in terms of popularity, but when there is lasting consensus regarding some concepts, we can assume that this reflects their functional relevance, at least seemingly. Having said that, we will try to show that, in life sciences *largo sensu*, the issue of using the word "engineering" is still problematic and has heavy consequences. This is because on the one hand, it seems that the notion of engineering is not neutral, in the sense that it conveys a whole series of theoretical assumptions; on the other hand, it is performative: it seems to not only describe but also often modify the understanding of the phenomena described in light of its original meaning. For example, as a first approximation, if it is tempting, to approximate a cell to a machine, then it is an entirely different matter to consider that, because it is a machine, a cell should then have certain properties, in particular, the repeatable accuracy of its mechanisms. The notion of engineering, which is implicit in the notion of machines and often without the knowledge of its users, allows such shift in meanings, and is therefore not a simple convenience or ease of language, but indeed a source of transformation of our own view of the living world. It is then important to check if this term removes some restrictions and reveals different realities or if, on the contrary, it alienates. This leads to wonder, in the evolutionary perspective specific to biology, whether the assimilation of its subjects (organisms, species and ecosystems) to machines or to technical or even industrial processes resolves more problems than it creates.

2.2. The animal–machine legacy

The Merriam-Webster Dictionary defines *engineering* as *"the application of science and mathematics by which the properties of matter and the sources of energy in nature are made useful to people"*. This defines a framework where engineering products are technical objects, derived from intentional and rational design. It seems therefore surprising that productions of the living world can be engineered, as they are, even if considerably modified when exiting a laboratory, entities characterized by a behavioral unpredictability and an evolutionary principle, which seems to radically differentiate them from machines that are intentionally created by humankind and whose sought characteristic is its reliability to execute tasks. Thus, it is through the expressions, life engineering and genetic engineering, that the idea that our species achieved a degree of control over such biological systems is implicitly conveyed and that biological systems could be subsumed by the notion of machines and related production methods. These expressions are then not neutral, in the sense that even if their use is now trivialized, they imply a certain form of implied pride linked to the performance they describe. In other words, talking about life engineering is simultaneously using this concept *and* implicitly opposing it to a previous time when this approach was out of reach. Yet, life engineering has its own history, which, in every era, was always presented as overcoming the challenges that biological systems pose to any rationalization attempt. Empirical, and then scientific, medicine was one illustration, but agronomy is probably the most striking example, particularly in the field of animal sciences, which were long known under the term zootechnics, which is in itself the assertion that animals could and even should be studied as machines. In fact, it is in this breeding context that the term "biotechnology" was coined [ERE 19]. In this sense, life engineering is an embodiment of a key concept of modern thought, the idea of animal–machine (and its expansion to living world–machine) inherited from René Descartes [DES 37].

It would be completely presumptuous to claim here to be summarizing the countless repercussions and implications of this concept, whose shock wave can still be felt in all sectors of research on living organisms and in the way society considers and gives a place to living organisms. However, we can nevertheless highlight a form of tension between what is liberating in this idea for science - namely the assertion that a living being, however complex, can be studied thanks to natural (and mechanical, in particular) laws, therefore without adding magical life forces and, on the contrary, how reductionist this idea is, namely that every living organism could be reduced to a machinist description or, in other words, to the product of bioengineering (even without any engineer other than nature). This questionable reduction comes from the confusion between mechanism and mechanization, which are actually two different things, as the Universe did not wait for the appearance of machines to be governed by the laws of mechanics. In addition, the commonplace assimilating of the living world and machines is contradicted and crushed by argumentation derived from several strong critical traditions, anchored in philosophy of science, such as the organicism of Georges Canguilhem [CAN 65], or in the more contemporary theoretical biology, such as the theory of Robert Rosen [ROS 91] who, with precision, differentiated the living world from machines, in so far as the first one "internalizes what ensures the permanence of its material system" [LEC 15] (what Rosen calls "closure to efficient cause"), while the second ones always depend on external maintenance. These theoretical attacks against living world–machines reinforces the idea that one can claim to materialistically analyze the living world, without reducing it to engineering, even natural engineering.

In fact, the living world naturally has its own mechanistic regularities, which partly explains why the similarities of individuals within a taxonomic group are more significant than those with external members: it contributes to what is called the biological order. It is one of the tasks of the biologist to explain this and the approximation of the living world–machine can contribute to it. However, besides that, the latter nevertheless far from includes the entire living world. It overlooks its multi-level structure (from biomolecules to ecosystems) and its specific complexity, even though the living world–machine advocates that nothing prevents these two aspects from being formalized, in the end, with engineering tools, especially thanks to the reinforcement of modeling approaches. However, the living world is really unamenable to any standardization when it comes to deploying it in several other dimensions: its historicity, fragility, unpredictability, sources of disorder, collective interactions, exploratory creativity and, of course, evolvability, which are all dimensions that are difficult to access with any form of bioengineering in the sense of rational planning and nearly completely missed by reducing any living organism to precision machinery.

2.3. Genetic engineering: rational design or tinkering?

Let us now take genetic engineering in particular as an example. It seems that, far from its apparently trivialized contemporary use, this notion was previously an incendiary topic. A rapid search in the PubMed database indicates that the oldest reference for the expression "genetic engineering" that can be found is probably in a text by one of the founders of molecular biology, the biochemist Rollin Hotchkiss, *Portents for a genetic engineering* [HOT 65], who attempted to assess the opportunities and risks of this approach, which he "prefers to call intervention", doubting in the process that one could ever talk about engineering in its strictest sense. It is found in a text of the evolutionist Théodosius Dobzhansky on the modification of humankind [DOB 67] and then in an article on living world engineering ethics [GOL 68]. Thus, during all these early years, concerns about ethics and in particular the risk of eugenics were omnipresent in the texts of this corpus. From a technical viewpoint, we find the phenomenon outlined earlier: it seems that, in every era, genetic engineering was always presented in opposition to previous techniques, which were reduced to mere tinkering, or even in opposition to the living world tinkering itself. This double shift deserves to be highlighted for several reasons. First, because it implicitly depicts a type of continuous representation between what living organisms spontaneously do and what *Homo sapiens* do themselves: there was a somewhat gentle transition between the spontaneous tinkering of nature and human tinkering on nature, which itself led to modern bioengineering. But mostly, it highlights the relativity of what belongs or does not belong to the engineering category: the first bacterial genetic transformations of the 1960s and the 1970s were seen as the advanced section of genetic engineering, whereas in 2017, they are now compared head-on with the most recent genetic engineering techniques, such as, for example, the use of the enzymatic system Crispr/Cas9 [CHA 13], which is now called "genome editing" to better state the accuracy expected from it, in order to modify genetic sequences to the nearest basis. Yet, if Crispr/Cas9 facilitates a resounding step forward in the accuracy to target the genetic modifications desired, there are still numerous gray areas (see, for example, Schaefer *et al.* [SCH 17]) on several intermediary stages between this targeting and its expected effect (for example, a high variability in the effectiveness of the transformation or in the effectiveness of the expression of the introduced sequence), highlighting at this early stage the perfectibility of this molecular tool. This portends interesting debates, in the next few years, on the amateurism that was in fact Crispr/Cas9, as opposed to the techniques replacing it. This downgrading at regular intervals can be relieving to us, but encourages, nevertheless, vigilance: if it can be understood that a technical concept changes as experimental progress is made, it is more problematic that what was used to support it is quickly considered as the opposite of what should define it.

2.4. Synthetic biology as the paradigm of bioengineering

The grand narrative of the last few years, which illustrates well this paradox, is that of synthetic biology. This term appeared as a multidisciplinary research program at the beginning of the 21st Century (for a detailed review, see Heams [HEA 09]), when the convergence of genome technologies, especially those of sequencing and DNA synthesis, with computer sciences and those of engineering, in particular electronic engineering, created a situation favorable to the idea that a genuine rational and machinist approach of the living world was within reach of experimentation. The radical transformation of the living world was in sight (new genes, new molecules supporting genetic information), and even life synthesis in the laboratory[1], at the end of a planned strategy to build increasingly complex modules. The context depicted in the founding articles of synthetic biology was that of the inexorable and imminent exhaustion of traditional approaches in biology (from naturalist observation to genetic tinkering), at the benefit of a new range of potential interventions, based on the study *a priori* of the logic of biological (genetic and/or metabolic) regulating circuits and their modeling, helping to rationally interfere with organisms, not by inserting a gene randomly, but accurately integrated, entire modules of genes, with controlled network effects: it is indeed a paradigmatic bioengineering that is asserted here, in that it implies a previous rational design, as opposed to any form of trial and error. Here, what was recalled earlier about the living world–machine all makes sense, because the logic of synthetic biology relies on the absolute equating of cells and precision machines, at the expense of the numerous dead ends previously mentioned.

Yet, this epistemological impoverishment has very practical consequences. To illustrate it, we can take as example, in this narrative about synthetic biology, the place occupied by an increasingly studied phenomenon, that of the intracellular random variation. A small detour is required here. One of the most important achievements in the last decades of cell biology is the experimental identification of random phenomena in intracellular function, particularly gene expression: it is now taken for granted that genetically identical cells can have non-deterministic behaviors, which are individually unpredictable, in a common and yet very homogeneous environment. If the very nature of this chaos can be argued, it nevertheless seems now proven that what happens in each cell does not have the metronomic regularity of the regulatory cascades described in biology manuals. We still need to interpret this observation, namely to give it a biological meaning, and it is here that our detour is justified. Some, for example, asserted that this unpredictability allows cells to have, under certain circumstances, an exploratory behavior and some flexibility to react, which can be an advantage [KUP 08]. It is a rich hypothesis, which can be

---

1 It is enlightening in this respect that one of the earliest articles about genetic engineering was also dedicated to the project of life synthesis in laboratory: *Progress in research related to genetic engineering and life synthesis [WID 74].*

scientifically tested and in fact seems to be increasingly strongly confirmed [RIC 16]. However, what synthetic biologists and other bioengineers say about it (although some of them also helped to highlight this random variation) is very different: in the living world–machine logic, this random variation is only cellular background noise for them, some kind of error margin in these living machines, whose accuracy is not yet perfectly smoothed out by natural selection (see, for example, Silva-Rocha & de Lorenzo, [SIL 10]). From a bioengineering perspective, this background noise must be controlled and, if possible, reduced, in order to ensure the reliability of the operation of these elementary machines. The fact that the random variation in the gene expression, that is background noise, is, after all, also has a possible hypothesis, even though it is increasingly contradicted by experimental indices. However, what is important is precisely that, as a hypothesis, it does not rely on its experimental verification, but on a very dogmatic assumption that it legitimates, an assumption according to which the living world could only be accurate and ordered: if traces of inaccuracy are identified, then they are necessarily negative and/or residual expressions that are intended to disappear. Far from a revelation, the engineering approach complicates the possible.

2.5. Transforming a transformation?

This way of thinking is in fact not new, because it has been accompanying molecular biology since its creation in the middle of the 20th-Century, which was based for a long time on the notion of *genetic program.* This term, borrowed from IT, which was emerging around the same time, is here again a sign of the long companionship between the living world and machines, where the computer is only the latest version of the metaphor, thus chronologically following hydraulic systems, steam engines, calculation machines, mechanical engines and factories, in the symbolic universe that has been shaping the representations of biologists for several centuries. Besides the already mentioned aporia of this metaphor, it is now necessary to insist on the aspect that is probably the most problematic one. The bioengineering rhetoric relies on a static vision of its subject to base its research program: it is about *transforming* the living world to better understand it. Yet, to say that is to assume its passivity because, precisely, *the living world is already a permanent transformation,* at all levels of space and time, from metabolic exchanges to the evolution of species. Understanding the living world is mainly, and with an essential humility, navigating between all the sources and dynamics of these changes in order to patiently expose some permanent and some regular aspects. Identity, in biology, is fundamentally a fable, even if, with great prudence, it can be legitimate to accept some of its expressions in order to better understand the mechanisms leading to it, the adaptive pressures and channeling and the converging phenomena. However, these regular aspects are fragile truths or otherwise very rare (for example, the universality on Earth of nucleic acids as hereditary material and the near-universality of the genetic code). Wishing to transform the living world as one would upgrade a computer that is cut off from the world apart from its network connection and its power supply is, strictly speaking, the same as wishing to *transform a transformation*, which might not be unthinkable but would require conceptual tools far more sophisticated than engineering for inert systems. This will not actually prevent synthetic biology from being successful, but can already help to explain why, for almost 20 years of effort, this claimed engineering has struggled to bring the revolution it promised. In addition, it is to be expected that one can play again the same score at other levels, because the popularity of engineering is not limited to the intracellular level. This is reflected by animal uplifting projects[2] (the expression comes from science fiction and refers to a "product line upgrade") or, the opposite, animal disenhancement [THO 08] aiming at modifying animals, especially breeding animals, like machines whose cognitive capacities could be respectively increased with neuronal grafts for example or, conversely, some unproductive organs could be completely eliminated through genome modifications. These projects, which seem to come straight out of sinister science fiction, nevertheless have promoters. They obviously raise ethical issues (animal welfare, subjection of living beings to human appetites for profit), and it is often the latter that debates focus on when the public learns about it. What is much less often highlighted, although this chapter is clearly an attempt to do so, is the epistemological dead ends of their foundations, which are however much more factual than ethical controversies, which are inevitably indexed in value systems that are never really universal. Highlighting these faults provides tools to demonstrate that, prior to being scientific rhetoric, these projects and representations are mostly symptoms of what a machinist vision of the living world can create as a fantasy. It then results from it that the engineering techno-reduction of the living world, far more than a fact or even a hypothesis, is also potential evidence of a strategy to muddy the

---

2 https://ieet.org/index.php/IEET2/more/cascio20141204

waters between scientific rigor and the brewing biopolitical ideologies, such as transhumanism and animal uplifting as its draft, for example.

2.6. Our cousins, the engineers

From one level to the other, the notion of engineering acquired an ecological purpose: concepts of engineer species and ecosystem engineering were proposed in 1994 by Clive Jones in an article, which immediately attracted the interest of ecologists, as evidenced by the significant number of times it has been cited [JON 94]. The concept of engineer species is based on the idea that some species transform their environment in a specific way by leaving lasting, indirect and significant evidence of their presence, which significantly modifies the resource flows and living conditions of other species. A canonical example is the beaver, which builds dams, or the termite community, but cases are varied and transversal in the living world, up to corals, which can modify currents and provide specific shelters to other marine species. After his founding article, ecosystem engineering was described as a typology of several cases, each illustrating a specific way for a species to be an agent transforming its environment, within the meaning of the definition. This approach makes it possible to base, in an experimental approach, predictions in the way some species, according to their specific characteristics, starting with their size and weight, can have a lasting impact on their ecosystem. Regardless of the intrinsic value of this concept, we wish however to critically discuss the use of the term "engineering" in this context. Since its origin, the term has in fact offended people, including within the ecologist community, for a rather obvious reason: the notion of engineering implies, in its strict sense, a form of intent doubled with a rational planning of the latter. Yet, if this can after all be legitimately considered in the case of a beaver dam, it seems irrelevant for corals. The founders of this approach are clearly aware of this ambiguity, and, in fact, they removed it very early by reclaiming a definition of engineering, leaving aside the issue of intent and focusing on actual effects and its functional aspect [JON 97]. This early clarification, which is in fact partly expressed in the founding article, helps to prevent several sterile semantic debates. Let us also highlight that this extension of the notion of engineering to other species, including beyond the *Animalia* taxon, also has the merit of naturalizing this practice, which is often considered as exclusively human, by detecting early stages of it somewhere else and by linking it to behaviors even in distant cousins in the tree of the living world. This makes it possible to remove from this notion an alleged aura, which would make it an invention specific to our species. As engineering is *a priori* consubstantial with a form of rationality, its use which is here extended is then stimulating, as it encourages us to seek the distant biological foundations of this rationality, against the comfortable hypothesis of its human exclusivity. However, beyond this valuable exercise and regardless of the usefulness that the notion can have for ecologists, we can nevertheless think that using the term in this context raises sensitive epistemological issues: if engineering is the capacity to leave a lasting indirect material evidence, even if it is unintentional and inevitably arising from a structure (corals do not have the ability not to create shelters), then everything or almost everything can be considered as engineering, because all living organisms have indirect cross-effects. This entails the risk of devaluing the notion itself because of this inconsiderate extension in all its other uses. Here, the issue is not that of its disciplinary usefulness, but of the revealing effect that this under-determined use provides to anyone who wonders about the biases we use to represent the living world in its multi-level entirety. In other words, the collective decision, within a research community, to use this term rather than another, reveals in fact, in an allegedly objective and neutral way, a hidden subjectivity, an unacknowledged projection of our machinist models and our desire to control what we call the living world. However paradoxical this can be, given that ecosystems, just like organisms and unlike machines, are also the result of permanent evolution.

2.7. Engineering and evolutionary dynamics

In order to put into perspective the effects that this vision of engineering, from the cell to the ecosystem, can generate, it is worth comparing it to evolutionary dynamics. To do so, we are going to temporarily return to synthetic biology, by highlighting the successive forms of its own actors' rhetoric on evolutionary dynamics. As highlighted, the synthetic biology of the 2000s was presented as engineers taking control of the living world, in particular as electronic circuit specialists. In the realm of the alleged gene networks described by molecular biology, these complex network specialists appear to be (self)designated to finally raise the hood of the living world and replace the circuits derived from an erratic Darwinian evolution with those derived from their *a priori* rational approaches (see, for example, Lazebnik, [LAZ 02]). Thanks to an accurate inventory of all the genes and control modules, the time would eventually come when the control of their precise and intentional assembly would outdo the blind operational evolution performances at work until then. Yet, synthetic biology was quickly faced with limitations in its results, in relation to its initial ambitions: *inter alia,* life was not and has never yet been created "from scratch" in a laboratory, and completely rewritten organisms with new codes or new hereditary molecules are still difficult to produce, not to mention their expected industrial applications. As the reasons for this resistance lie in the numerous black boxes of the living world, in particular the cross-effects of genes on each other and their environmental interactions, it is then that Darwinian dynamics became interesting again [COB 12], this time in the form of guided evolution, *in vitro* (to select molecules of interest) or *in vivo* (to select cell lines adapted to constraints). This technique, based on the variation/selection principle applied in a laboratory through a control of selection constraints and pressures, proved to be very useful to unblock the attachment points upon which the rational design, *a priori* dear to synthetic biology engineers, stumble over. Thus, these dynamics, which imitate the driving force of biological evolution, but were initially looked down on by synthetic biologists, demonstrate that engineering, as it had been initially proclaimed as prior rational planning, only deserved a relative place in the range of tools that made it possible to understand and transform the living world. And it is even more often on random-selection dynamics that the most efficient tools of genetic transformation rely nowadays [WAN 09].

Will we witness similar shifting movements in other disciplines of the living world and environmental sciences? Several clues suggest it, because once again we are now at the dawn of a convergence of these approaches. Gene drive, for example, a technique derived from Crispr/Cas9 approaches, promises to increase the speed with which genes can spread in a population, for example, by cultivated plants or, conversely, by insects bearing diseases. A multi-level combined engineering, from cell to ecosystem, is then considered by its promoters in order to modify the environment for our use, at the risk of underestimating its poorly controlled consequences and the evolutionary consequences on the biosphere. Another example of bio-techno-ecological convergence is de-extinction projects [SHA 15], which promise to bring back to life (*sic*) extinct species, thanks to genetic engineering on genomes of closely related species, and which is also a rewilding project (see [DON 05]), namely the colonization of entire natural areas by these resurrected species. Apart from their questionable objective and also mainly the degree of doubtful realism of each step taken, beyond the meditation they encourage on the role of self-fulfilled prophesies in biology, we see that there is a temptation of hubris, where the wild world is under control and where nothing is prohibited, because as engineers, we would control each section of a problem: strange ethics, which would dictate that because we control it, goals would be per se desirable. Here, the danger of getting everything mixed up and using imperfectly delimited concepts-bridges is materialized in the abuses and unfounded perspectives it conjures. So here, we can gauge the risk of seeing the living world only through the prism of engineering at all levels.

In fact, other paths remain. The one that would actually make us realize the resilience of biological systems despite our interventions with biotechnological weapons, where biotechnology would sometimes be an operational tool rather than the solution, but most of the time would mainly allow us to understand how inventive ecosystems are and how they throw off predictions. The one that would make us learn the best lesson through facts and ecosystem engineering, namely the prudence needed in our interventionist, and even disturbing, approaches, the required trial and error, the patience to experiment on a range of increasing levels before making predictions on the overall value of local models. These paths would then maybe indicate that what we call bioengineering and the inevitable claim of control it conveys, which is illustrated by Ridley Scott's Engineers, is above all a very relative, temporary and reversible domestication of some living organisms. It is a less dashing description, but maybe more realistic and therefore more productive of what we, as a species, freshly contributed to the transformation of living organisms by each other. In addition, it is also in some way a contribution in order to define our rightful place in the biosphere, a prerequisite for better understanding our responsibilities. Figuring the living world by getting rid of engineering could then be, for several reasons, a necessity.

2.8. Bibliography

[CAN 65] CANGUILHEM G., *La Connaissance de la vie*, Paris, Vrin, 1965.

[CHA 13] CHARPENTIER E., DOUDNA J.A., "Biotechnology: Rewriting a genome", *Nature*, vol. 7, no. 495(7439), pp. 50–51, March 2013.

[COB 12] COBB R.E., SI T., ZHAO H., "Directed evolution: an evolving and enabling synthetic biology tool", *Current Opinion in Chemical Biology*, vol. 16, nos 3–4, pp. 285–291, August 2012.

[DES 37] DESCARTES R., *Discours de la Méthode, Livre Cinquième*, Jan Maire, Leiden, 1637.

[DOB 67] DOBZHANSKY T., "Changing Man", *Science*, vol. 27, no. 155(3761), pp. 409–415, January 1967.

[DON 05] DONLAN J., "Re-wilding North America", *Nature*, vol. 18, no. 436(7053), pp. 913–914, August 2005.

[ERE 19] EREKY K., *Biotechnologie der Fleisch-, Fett-, und Milcherzeugung im landwirtschaftlichen Grossbetriebe: für naturwissenschaftlich gebildete Landwirte verfasst*, Paul Parey, Berlin, 1919.

[GOL 68] GOLDING M.P., "Ethical issues in biological engineering", *UCLA Law Review*, vol. 15, no. 2, pp. 443–479, February 1968.

[HEA 09] HEAMS T., *De quoi la biologie synthétique est-elle le nom ?*, in HEAMS T. et al., *Les Mondes Darwiniens*, Editions Matériologiques, Paris, 2009.

[HOT 65] HOTCHKISS R.D., "Portents for a genetic engineering", *Journal of Heredity*, vol. 56, no. 5, pp. 197–202, September–October 1965.

[JON 94] JONES C.G., LAWTON J.H., SHACHAK M., *Organisms as Ecosystem Engineers Oikos*, vol. 69, no. 3, pp. 373–386, 1994.

[JON 97] JONES C.G., LAWTON J.H., SHACHAK M., "Ecosystem engineering by organisms: why semantics matters", *Trends in Ecology & Evolution*, vol. 12, no. 7, p. 275, July 1997.

[KUP 08] KUPIEC J.J., *L'origine des individus*, Fayard, Paris, 2008.

[LAZ 02] LAZEBNIK Y., "Can a biologist fix a radio? – Or, what I learned while studying apoptosis", *Cancer Cell*, vol. 2, no. 3, pp. 179–182, September 2002.

[LEC 15] LECHERMEIER G., Définition du vivant et émergence de la vie : entre rupture et continuité, saisir l'originalité du vivant, PhD thesis, Pantheon-Sorbonne University, 2015.

[RIC 16] RICHARD A., BOULLU L., HERBACH U. et al., "Single-cell-based analysis highlights a surge in cell-to-cell molecular variability preceding irreversible commitment in a differentiation process", *PLoS Biology*, vol. 14, no. 12, p. e1002585, 2016.

[ROS 91] ROSEN R., *Life Itself: A Comprehensive Inquiry into the Nature, Origin, and Fabrication of Life*, Columbia University Press, New York, 1991.

[SCH 17] SCHAEFER K.A., WU W.H., COLGAN D.F. et al., "Unexpected mutations after CRISPR-Cas9 editing in vivo", *Nature Methods*, vol. 30, no. 14(6), pp. 547–548, 2017.

[SHA 15] SHAPIRO B., *How to clone a Mammoth?*, Princeton University Press, United States, 2015.

[SIL 10] SILVA-ROCHA R., DE LORENZO V., "Noise and robustness in prokaryotic regulatory networks", *Annual Review of Microbiology*, vol. 64, pp. 257–275, 2010.

[THO 08] THOMPSON P., "The opposite of enhancement: nanotechnology and the blind chicken problem", *Nanoethics*, vols 2–3, pp. 305–316, 2008.

[WAN 09] WANG H.H., ISAACS F.J., CARR P.A. *et al.*, "Programming cells by multiplex genome engineering and accelerated evolution", *Nature*, vol. 13, no. 460(7257), pp. 894–898, August 2009.

[WID 74] WIDDUS R., AULT CR., "Progress in research related to genetic engineering and life synthesis", *International Review of Cytology*, vol. 38, pp. 7–66, 1974.

3

The View of Systematics on Biodiversity

3.1. Introduction

There are several ways to evaluate Biodiversity, which is a multi-level concept with multiple purposes. Each consideration of this concept generally involves the prism of a scientific discipline, a study level or a specific purpose [MAC 08]. Thus, scientists focus sometimes on intraspecific diversity, sometimes on interspecific diversity, or even on the diversity that can be studied at the level of ecosystems, etc. Disciplines usually concentrate on their study subject, such as genomics on diversity within genomes or population biology on diversity within populations.

For my part, I would like to share several general reflections on Biodiversity, which are inspired by the scientific contribution of a discipline, Systematics. The origin of this discipline is ancient and is still often assimilated in people's minds with description and classification tasks, as they were practiced at the time of the great naturalists of the 18th and 19th Centuries. However, for over half a century, Systematics has been a major player in evolution biology, in particular with phylogenetic analysis, which appeared after the work of the systematist Willi Hennig in the 1950s 1960s and which spread throughout Biology [OHA 92].

Systematics' contribution pertains to the field of comparative biology [NEL 70]: it compares organisms and their characteristics, and draws

Chapter written by Philippe GRANDCOLAS.

conclusions in terms of evolutionary relationships (phylogenies, homologies); it establishes diagnoses (taxonomy). Far from denying the variability of the living world [DEB 01], which is especially integrated in species concepts, it studies it, on the contrary, to select invariants, which will make it possible to study the origin of Biodiversity, namely the way organism characters are set up.

My goal is not to present Systematic case studies of biodiversity, but to highlight several significant points which Systematics can help us to understand or value regarding Biodiversity, and which are linked in a very interesting way to those uncovered by ecological approaches. The purpose of this presentation is obviously not to promote a discipline for its own sake, but to provoke reactions and reflections through the contrast between different points of views.

3.2. Species: all different

In the 1980s, the creation of the concept of Biodiversity and the advent of phylogenetic analysis made it possible to reintroduce in people's minds a very valuable asset: the notion of diversity between species, which had faded away since the rise of General Biology[1] at the beginning of the 20th Century [GRA 17].

In fact, General Biology and what we still call today Life Sciences (with some components called Systems Biology[2]or Integrative Biology[3]) appeared and developed with the idea of seeking and defining main general principles common to all organisms, like, for example, the "laws of heredity". It is also at this time that model organisms, which were supposed to represent on their own entire sections of the living world, emerged, such as the Thale Cress *Arabidopsis thaliana* for plants or the common fruit fly *Drosophila melanogaster* for insects. From the moment we focus on the main constitution or functioning principles supposedly common to all organisms, we are concerned far less with their differences and diversity. General

---

1 General biology studies mechanisms in a few species in detail and draws conclusions that can supposedly be generalized to the functioning of the whole living world.
2 Systems biology studies the functioning of living systems from the cell to the tissue, by focusing on cellular and molecular approaches.
3 Integrative biology integrates physiological and biophysical aspects in the study of organism functioning.

Biology thus contributes to the study of organism evolution, because it identifies general heredity or functioning mechanisms of organisms. However, it does not then study the evolutionary History of different groups of organisms in interaction with the environment, which is the source of Biodiversity.

This eclipse of History (evolutionary History), as the systematist Dan Brooks and the ethologist Deborah McLennan called it [BRO 91], then ended in the 1980s when the community of biologists became aware again of the importance of considering differences between species. At that time, the term biodiversity was coined by Thomas Lovejoy, Walter G. Rosen and Edward Osborne Wilson [WIL 88]. Phylogenetic analysis was formalized at the same time [WIL 81], which allowed it to benefit from the advent of molecular biology and its massive production of data to study Biodiversity. The study of the living world thus regained not only intraspecific variation (population genetics in particular) and interspecific interactions (ecology), the study of which was well developed since the beginning of the 20th Century, but also differences between species, the study of which was less advanced, especially at the evolutionary level. In all these ways, differences between species again became a subject of analysis and not a background noise to be removed in order to study the laws of the living world.

This awareness that the variety of living organisms must be a subject of scientific and societal concern, as much as the unicity of some processes in the whole living world, was reflected in many different ways, including Biodiversity sciences. For example, Biodiversity metrics changed, especially thanks to Systematics. We went from the rather universal use of specific richness, a traditional measure in ecology that considers all species as identical (a tribute to General Biology at the beginning of the 20th Century), to more sophisticated metrics.

This way, Faith [FAI 92] offered phylogenetic diversity or PD, a measure taking into account, for a set of individuals, the total length of the phylogenetic tree branches connecting them: it thus expresses the quantity of common characteristics and differences characterizing them. Even though it is correlated with the specific richness (the more species are considered, the more the number of branches linking individuals in a tree, and therefore the sum of their lengths), PD provides much more information than a simple cardinal sum. It has the great advantage of being a potentially overall measure of genetic and phenotypic diversity, and thus goes beyond geographic, regional or local levels [VER 15, PEL 16a].

We could mention another example with the first version of the "comparative method" of the 1980s, which aimed to calculate alleged statistical causal relations between traits and environments [CLU 79]. This method considered species as equivalent once again and simply controlled that they were not statistically independent in cases of close relationships (comparing ten squirrels and ten mice would statistically come to only comparing two groups!).

This method then evolved into the phylogenetic analysis of evolution (with modern "comparative phylogenetic methods"), which compares several species' evolutionary histories immediately considered as different, histories that are rebuilt by establishing potential homologies [BRO 91].

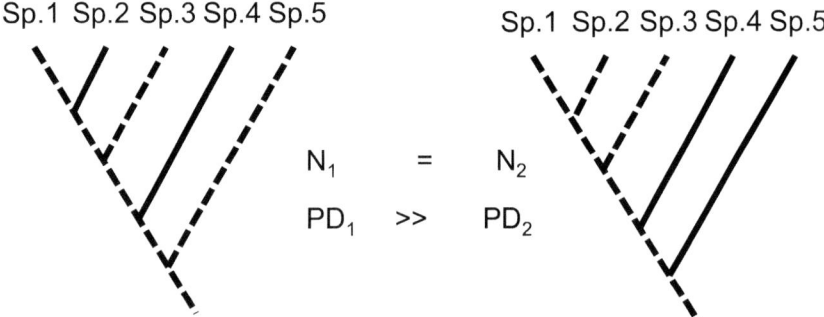

Figure 3.1. *A theoretical phylogenetic tree in which two sets of three species are chosen (dotted lines). The specific richness of the two sets is the same (N=3), but the PD of the left selection is much higher (especially with sp.5 which is the sister-group of the rest of the tree instead of sp.2, the sister-group of sp.1)*

3.3. How about studying the other 90%?

This welcome awareness of the significance of the variety of living organisms certainly also helped us to become fully conscious of the terrifying disaster represented by the current crisis of Biodiversity. The Biodiversity crisis is serious, because individuals belonging to different species represent different values, functions and services: reducing their number and variety is equivalent to direct losses in all these aspects.

Nevertheless, there is still an intellectual frontier that Systematics can help us to cross, understanding the significance of the 90 % of current Biodiversity that is still unknown. It is in fact a paradox to read the alarming numbers of extinctions, which are unfortunately already effective or to come within a few decades [BAR 11, REG 15] and which apply to what we know (10%), and which unintentionally bypass what we do not know (90%). Without being particularly concerned with cataloguing or inventorying, we should still acknowledge that our sample of the living world is hardly representative in light of the numerous issues raised, including the crucial issue of the extinction crisis.

Thus, do we really know what is going to disappear and, *a fortiori*, what was the role of the biodiversity that disappeared before we could know it [REG 09]? Can we be content with the possibility of a probable functional redundancy essential to the functioning of ecosystems, without really knowing how many species are concerned and really redundant? From this point of view, and knowing that rare species can play significant roles in ecosystems [MOU 13], which serious consequences are the extinctions of rare species that are still unknown going to have? What would the 90% remaining Biodiversity provide in terms of knowledge, especially in the very current field of Biomimetics or Bioinspiration [BEN 16]? For the record, out of approximately one million known species, only 76,000 are concerned by patents [OLD 13]. What would it be with ten times more species known?

In view of this paradox - extinction of a mostly unknown set - the most common response is often only operational or logistic: we do not know enough about Biodiversity, never mind, let us use a particular tool or procedure and, in 20 years, we will have substantially increased our knowledge [MAY 04, DAL 12]. However, without neglecting the main programs of Biodiversity exploration, to which we can only subscribe with enthusiasm, it seems mainly important to realize that we should all address the issue of the still unknown Biodiversity and not leave it to the exploration programs of the living world alone (Systematics Agenda 2000, Barcode of Life, *La Planète revisitée*, and soon Planetary Biodiversity Mission), almost as if the genomics discipline had only been built and practiced by consortiums composed for the sequencing of the first complete genomes.

This means that there is a requirement to improve ethics and to make taxonomic knowledge available, the requirement that should be shared by all scientists studying Biodiversity. Let us not continue thinking in terms of general biology, as if knowledge research should only be content with the quest of universal aspects (laws, processes, etc.). We must realize that the increase in knowledge on Biodiversity specificities is also a key study subject. In this respect, the Nagoya Protocol and its national regulatory implementation are often seen as a constraint by scientists, whereas they helpfully remind us of our ethical obligations. Laboratories studying biodiversity must train or house scientists competent in taxonomy, who are nowadays disappearing [GRA 12]: it is not enough to barcode, metabarcode, database, digitalize, geolocalize, etc., specimens to save information specific to a study or useful to society. It is also necessary to directly contribute to taxonomy, by describing or revising taxa. Information and knowledge access and sharing are only possible through a taxonomic system with reference specimens and names. The amazing molecular, digital and computer means, which are rapidly developing, only make this linkage even more vital [PEL 16b]. This challenge is particularly crucial in the case of micro-organisms, whose accessibility is limited by our powers of perception, and which are important in all respects, including through their omnipresence within other organisms, with which their interactions (for example, mycorrhizae, intestinal microbiome, etc.) are essential [SEL 17]. In their case, molecular data are directly integrated into a traditional taxonomy, which is very much alive. Nevertheless, beyond metabarcoding approaches, the ability to cultivate species or keep strains is a significant issue that microbiology laboratories have well understood.

3.4. Biodiversity changes

Systematic and comparative design of Biodiversity also reminds us that it is derived from a long evolutionary process, that it can be phylogenetically characterized, including at a specific level, and that it is not given as a natural kind, in a sort of "fixist" or more specifically essentialist conception [ROB 17] as our short human perception level would suggest.

It is necessary nowadays to reconcile the so-called functionalist, "microevolutionist" and "macroevolutionist" perspectives. There is no useful and significant microevolutionist evolution biology considering the population mechanisms to which we are confronted at our timescale (a few years or a few decades), and another quaint and "cultural" macroevolutionist evolution biology dealing with fossils and building phylogenies[*] for

Platypus, Ginkgos and other Dodos at a timescale incompatible with our existence (millions of years).

We must know the origin and selective context of phenotypic traits to study them and understand the adaptive phenomenon at the population level. The great development of "tree-thinking" (that is, the use of phylogenetic trees) in biology since the 1990s has taught us that a lack of macroevolution contextualization often made us build the microevolutionary explicative model backwards [BRO 91]: for example, in some spiders, models trying to explain sexual dimorphism through male dwarfism strayed by not considering the gigantism of females [COD 97]. Not linking micro-and macroevolutionary studies would be like rejecting all the watches with no second hand, because we are disappointed not to see the movement of the hour and minute hands when quickly looking at the watch face. And yet, we need to tell the time.

This level of – systematic and phylogenetic – Biodiversity macroevolutionary study is not only an essential explanatory foundation to conduct microevolutionary studies, it also helps to assess and perceive the living world in all its diversity. Once more, phylogenetic diversity is not only a good overall metric, but it also allows us to ethically consider Biodiversity.

In a phylogenetic tree, no organism is superior to another. All current or fossilized organisms, although they are very different, are the leaves of a tree, the root of which is an abstraction: there are only sister-groups within a tree, which does not show genealogies, but the sister-group relationships between species [GRE 08, CRI 05, GRA 14]. Phylogenetic trees are thus appropriate media to explain that gradist[4] or anthropocentric reasoning is absurd, both for scientists and for all publics [FOR 09, MAC 12].

3.5. Challenging decades

We are thus at a crucial and paradoxical moment for the study and consideration of Biodiversity. It has never been so well known, but its

---

4 Misleading concept, the origin of which is often associated with Aristotle's Chain of Beings; it considers that the living world is graded according to complexity with the simplistic idea of an evolutionary progress that would result from adaptation; anthropocentrism is a form of gradism, because it puts human kind at the top of the grade.

greatest part still remains to be discovered. This whole biodiversity - known and unknown - is at risk of major extinction in the coming decades. Our means to study it have never been so powerful but, paradoxically, the flood of data created by these means is in itself a challenge for knowledge access and sharing. This challenge also has a strong geopolitical component. Biodiversity is particularly significant at low latitudes, and access and sharing must not only be an ethical concern common to the whole of humanity, but also a desire to rebalance knowledge, expertise and means between different political powers, in order to better share the environment common to all of us.

3.6. Bibliography

[BAR 11] BARNOSKY A.D., MATZKE N., TOMIYA S. *et al.*, "Has the Earth's sixth mass extinction already arrived?", *Nature*, vol. 471, pp. 51–57, 2011.

[BEN 16] BENYUS J.M., *Biomimétisme: Quand la nature inspire des innovations durables*, Rue de l'échiquier, Paris, 2016.

[BRO 91] BROOKS D.R., MCLENNAN D.A., *Phylogeny, ecology, and behavior: a research program in comparative biology*, The University of Chicago Press, Chicago, 1991.

[CLU 79] CLUTTON-BROCK T.H., HARVEY P.H., "Comparison and adaptation", *Proceedings of the Royal Society of London,* vol. 205, pp. 547–565, 1979.

[COD 97] CODDINGTON J.A., HORMIGA G., SCHARFF N., "Giant female or dwarf male spiders?", *Nature*, vol. 385, pp. 687–688, 1997.

[CRI 05] CRISP M.D., COOK L.G., "Do early branching lineages signify ancestral traits?", *Trends in Ecology & Evolution*, vol. 20, pp. 122–128, 2005.

[DAL 12] DALY M., HERENDEEN P.S., GURALNICK R.P. *et al.*, "Systematics agenda 2020: the mission evolves", *Systematic Biology*, vol. 61, pp. 549–552, 2012.

[DEB 01] DEBAT V., DAVID P., "Mapping phenotypes, canalization, plasticity and developmental stability", *Trends in Ecology & Evolution*, vol. 16, pp. 555–561, 2001.

[FAI 92] FAITH D.P., "Conservation evaluation and phylogenetic diversity", *Biological Conservation*, vol. 61, pp. 1–10, 1992.

[FOR 09] FORTIN C., LECOINTRE G., BÉNÉTEAU A., *Guide critique de l'évolution*, Belin, Paris, 2009.

[GRA 13] GRANDCOLAS P., DAUBIN V., CHAVE J. et al., "Systématique, Phylogénie", in THIÉBAULT S., HADI H. (eds), *Prospective de l'Institut Ecologie Environnement du CNRS. Compte-Rendu des Journées des 24 et 25 Octobre 2012, Avignon*, CNRS, Paris, 2013.

[GRA 14] GRANDCOLAS P., NATTIER R., TREWICK S.A., "Relict species: a relict concept?", *Trends in Ecology & Evolution*, vol. 29, pp. 655–663, 2014.

[GRA 17] GRANDCOLAS P., "Loosing the connection between the observation and the specimen: a by-product of the digital era or a trend inherited from general biology?", *Bionomina*, vol. 12, pp. 57–62, 2017.

[GRE 08] GREGORY T.R., "Understanding evolutionary trees", *Evolution: Education and Outreach*, vol. 1, pp. 121–137, 2008.

[MAC 08] MACLAURIN J., STERELNY K., *What is biodiversity?*, University of Chicago Press, Chicago, 2008.

[MAC 12] MACDONALD T., WILEY E.O., "Communicating phylogeny: evolutionary tree diagrams in museums", *Evolution: Education and Outreach*, vol. 5, pp. 14–28, 2012.

[MAY 04] MAY R.M., "Tomorrow's taxonomy: collecting new species in the field will remain the rate-limiting step", *Philosophical Transactions of the Royal Society of London*, vol. 359, pp. 733–734, 2004.

[MOU 13] MOUILLOT D., BELLWOOD D.R., BARALOTO C. et al., "Rare Species Support Vulnerable Functions in High-Diversity Ecosystems", *PLoS Biology*, vol. 11, p. e1001569, 2013.

[NEL 70] NELSON G.J., "Outline of a theory of comparative biology", *Systematic Zoology*, vol. 19, pp. 373–384, 1970.

[OHA 92] O'HARA R.J., "Telling the tree: narrative representation and the study of evolutionary history", *Biology and Philosophy*, vol. 7, pp. 135–160, 1992.

[OLD 13] OLDHAM P., HALL S., FORERO O., "Biological diversity in the patent system", *PLoSONE*, vol. 8, p. e78737, 2013.

[PEL 16a] PELLENS R., GRANDCOLAS P. (eds), *Biodiversity Conservation and Phylogenetic Systematics: preserving our evolutionary heritage in an extinction crisis*, Springer Open, Berlin, 2016.

[PEL 16b] PELLENS R., FAITH D.P., GRANDCOLAS P., "The future of phylogenetic systematics in conservation biology: linking biodiversity and society", in PELLENS R., GRANDCOLAS P. (eds) *Biodiversity Conservation and Phylogenetic Systematics: preserving our evolutionary heritage in an extinction crisis*, Springer Open, Berlin, 2016.

[RÉG 09] RÉGNIER C., FONTAINE B., BOUCHET P., "Not knowing, not recording, not listing: numerous unnoticed mollusk extinctions", *Conservation Biology*, vol. 23, pp. 1214–1221, 2009.

[RÉG 15] RÉGNIER C., ACHAZ G., LAMBERT A. *et al.*, "Mass extinction in poorly known taxa", *Proceedings of the National Academy of Sciences of the USA*, vol. 112, pp. 7761–7766, 2015.

[ROB 17] ROBERT A., FONTAINE C., VERON S. *et al.*, "Fixism and conservation science", *Conservation Biology*, vol. 31, pp. 781–788, 2017.

[SEL 17] SELOSSE M.A., *Jamais seul - Ces microbes qui construisent les plantes, les animaux et les civilisations*, ActeSud, Arles, 2017.

[VER 15] VERON S., DAVIES T.J., CADOTTE M.W. *et al.*, "Predicting loss of evolutionary history: Where are we?", *Biological Review*, vol. 92, pp. 271–291, 2015.

[WIL 81] WILEY E.O., *Phylogenetics, The theory and practice of phylogenetic systematics*, Wiley-Liss, New York, 1981.

[WIL 88] WILSON E.O. (ed.), *Biodiversity*, National Academies Press, Washington, 1988.

4

Which Model(s) Explain Biodiversity?

Like any natural science (for example, physics or chemistry), biology is characterized by a double approach combining observation and theory. Although these two aspects were historically conducted together by great scientific figures (let us mention, for example, Darwin, Galton, Mendel or Fisher for evolutionary biology), both approaches gradually separated during the 20th Century, thus creating two complementary communities. The first one gathered naturalists who mainly focused on the study of diversity *patterns,* and worked in order to describe and organize individuals and species into clades, which are the relevant unit for the phylogenetic classification of living organisms. The second one was rather concerned by the study of the underlying *processes* shaping biodiversity, in order to single out and model the main mechanisms at work in Nature. This separation greatly boosted these two branches and generated researchers who were expert in their field. Unfortunately, it also resulted in a decline in the dialogue between "experimental biologists" and "theoreticians", which only started again with the massive influx of genomic data, hard to analyze without the support of models.

In this chapter, we will study the mechanistic foundations of two evolutionary models, which are commonly used in connection with our research in evolutionary biology, by trying to provide some simple elements

Chapter written by Guillaume ACHAZ.

in order to understand their essence, by showing how they are used in data analyses and, finally, by discussing their limits in light of our knowledge in biology.

A clear separation between *macroevolutionary* processes governing species evolution and *microevolutionary* processes explaining diversity within one or a few species is very often accepted. This dichotomy probably derives its legitimacy from the hardly questionable observation that individuals within the same species are generally less different than those from different species. Yet, is this separation of taxonomic levels a sufficient justification for understanding interspecific diversity in a totally different way from intraspecific diversity? Or is it not time to delve at the very heart of these abstract descriptions in order to test their use and biological relevance?

Here, we will deal with two classes of models, respectively *birth–death* models [KEN 48], which explicitly model a genealogy built in the prospective sense of time, and *coalescent* models [KIN 82], in which the genealogy of individuals at the present time emerges from the retrospective analysis of the ancestry of a population of individuals.

Historically, *birth–death* models were used to describe the diversity of species, while *coalescent* models relate to population genetics: the diversity analysis within a species. However, the adoption of these evolutionary models in order to describe intra-individual cellular diversity in immunology [FRA 02] or even in cancer research [BEE 15] shows us that these models are general and applicable to organization levels of the living world other than those for which they were initially created.

One of the reasons for the popularity of *birth–death* models is their application to inferred phylogenies, in order to understand which processes created these phylogenies [NEE 94]. This approach is the same in population genetics, where polymorphism data analyses are used, apart from the description of kinships between individuals, to infer the mechanisms that generated this kinship [HAR 97]. Thus, in this school of thought, the tree is no longer a goal *per se* it is the intermediary that allows us to understand the evolutionary mechanisms reflecting the diversity observed (Figure 4.1).

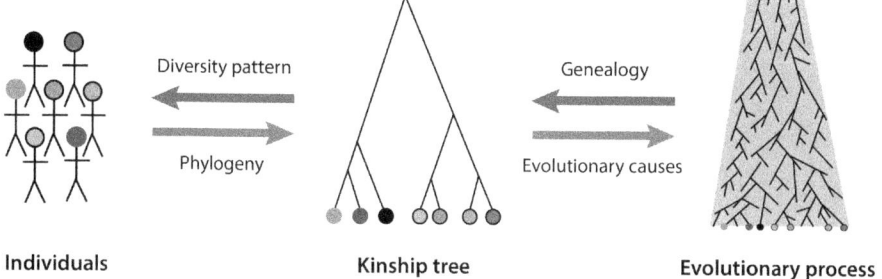

Figure 4.1. *Conceptual framework. From right to left, this illustrates how evolutionary processes generate kinship trees, which in turn reflect the diversity of a group of individuals or species. From left to right, the explicit use of quantified evolutionary models first makes it possible to infer this (or these) kinship tree(s), which are in turn used to understand the causes of their evolution. For a color version of this figure, see www.iste.co.uk/grandcolas/biodiversity.zip*

4.1. Birth–death processes

Birth–death models generate, assuming relatively simple hypotheses, stochastic trees, which are built in prospective time. An origin of the birth–death processes is found in Bienaymé–Galton–Watson processes [GAL 75], which describe in discrete time the size variation of a population composed of several individuals. At each generation, each individual gives rise to a random number of offspring composing the next generation. The filiation of each individual can be followed over time and the Galton–Watson tree can thus be obtained, which is a mathematical object dear to probabilists. Birth–death processes are a continuous equivalent to these discrete processes.

We only present below a few elements to better understand these birth–death models, in order to be able to use them with the critical eye required for a reasoned use. We invite the reader wishing to further explore this matter to read articles on the numerous avatars and their use in macroevolution (for example, [NEE 06, MOR 14]).

The first birth–death model was a pure birth model [YUL 24], in which each species is modeled as an *evolutionary lineage*. A lineage represents the succession of generations of individuals composing the species over time. It is typically represented by a line whose length corresponds to the existence time of this species. In the Yule model, each lineage gives rise to new lineages in exponential time (Figure 4.2(a)). It should be noted that an exponential time represents a waiting time during which an event (here speciation) can occur with an identical probability each time. This genesis process of new lineages can be assimilated to particle fissions which, from the model point of view, are strictly equivalent.

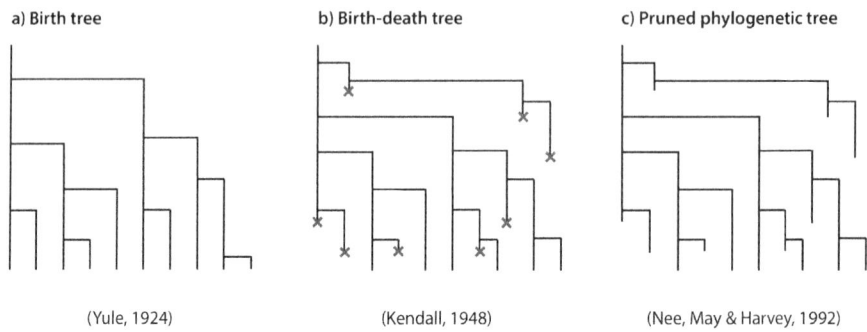

Figure 4.2. *Trees generated by birth–death processes. a) In the Yule model, only births are modeled and the total number of lineages increases exponentially. b) In the birth–death model, the diversification rate is provided by the difference between the birth and death rates. c) When a phylogeny is considered, only the lineages still alive at the present time are represented in the phylogeny*

In this abstraction, a species is represented by a simple lineage, ignoring all of the behaviors and specificities of the individuals composing it. Even though this simplification is hard to accept for a biologist, it is conceptually close to the physics tools (such as mean fields) used to describe a set of particles. In addition, speciation is assimilated with an instantaneous and irrevocable event. Even though this assertion is also clearly wrong, it is understood that when speciation times are extremely short in relation to species persistence, this speciation can effectively be assimilated with punctual events at the scale of interest.

In the Yule model, no death can occur and each lineage splits at a constant rate. The number of lineages alive after the foundation date increases exponentially up to infinity in a nearly determinist way. The growth speed of such a clade depends on the speed at which each lineage splits. When n lineages are alive, as each splits at the rate b, the probability of observing the birth of a $(n+1)^{th}$ lineage during a time dt is $n \times b \times dt$. The average waiting time is $1/nb$. The more lineages there are, the smaller the waiting time to observe the birth of a new lineage.

Yet, the number of lineages of a real clade is certainly not infinite. You should then imagine that either the speciation rate b is very low, or that it is counterbalanced by lineage extinctions. In a more general birth–death model, each lineage gives birth to new offspring at rate b (for *birth*), but can also die out at rate d (for *death*). If speciation and extinction are modeled as two independent exponential variables, the probability for a lineage to multiply before becoming extinct is provided by the ratio $b/(b+d)$. Thus, when b is very high before d, this probability tends towards 1. *A contrario*, when b is very low before d, this probability then tends towards 0 and all the clade lineages become extinct.

The birth–death model (Figure 4.2(b)) is, in this more general form, completely described by three parameters: a foundation date t_0, a speciation rate b and a death rate d [KEN 48]. Classically, three regimes for this model are identified. A first regime called *subcritical* corresponds to the case where $b<d$, namely where the extinction rate is superior to the speciation rate. The extinction probability of all the lineages alive is equal to one: a clade evolving in a subcritical regime is bound to disappear, if a researcher waits long enough. The speed at which this extinction occurs, increases with the d/b rate ratio, but also with the absolute value of d, which measures the extinction process speed of each lineage. It is the same for the *critical* regime where $b=d$, and where the average extinction speed increases with this single rate. The higher this rate is, the higher the number of speciation and extinctions per unit time and the faster the complete extinction of the clade. Only the *supercritical* regime, $b>d$, has a total extinction probability lower than 1 (even though it is not null). In this last case, when a sufficiently high number of lineages are reached, the number of lineages starts to increase exponentially, as in the pure birth model, but with a growth rate equal to b-d. This rate is called the *diversification rate* of the clade.

Interpreting actual data in light of birth–death stochastic processes reflects a paradigm change in paleontological studies. The study subject which had focused up until then on the search for causal explanations for various diversity variations – the so-called "idiographic" approach – was replaced by the search for a general framework helping to understand biodiversity variations as a whole – the so-called "nomothetic" approach [RAU 73]. This deep change in paleontology is the work of a small working group, which used computer simulations to explain diversity variations in fossil registers. These simulations are discrete birth–death processes [RAU 73, GOU 77], but with rate values of birth b_t and death d_t that vary over time. More specifically, they vary according to the difference between the number of lineages alive L_t at the time t, and a constant K, which represents a number of lineages corresponding to a diversity equilibrium. K is, in a way, a *carrying capacity* of diversity. When $L_t{>}K$, the simulator adjusts rates, such as $d_t{>}b_t$, and the reverse. Thus, the number of lineages, which is instantiated at one at the beginning, increases until it reaches the value K, around which it oscillates indefinitely. Except for the few oscillations, this second phase corresponds to a model with a constant size that we will describe later. We will then show the simulator outputs, which really look like the variations in the paleontological number of fossils over time (Figures 4.3(a) and 4.3(b)). Driven by the desire to explain paleontological data using ever simpler processes, the diversity saturation phase was then eliminated, and the birth and death rates reset to constant values [STA 81].

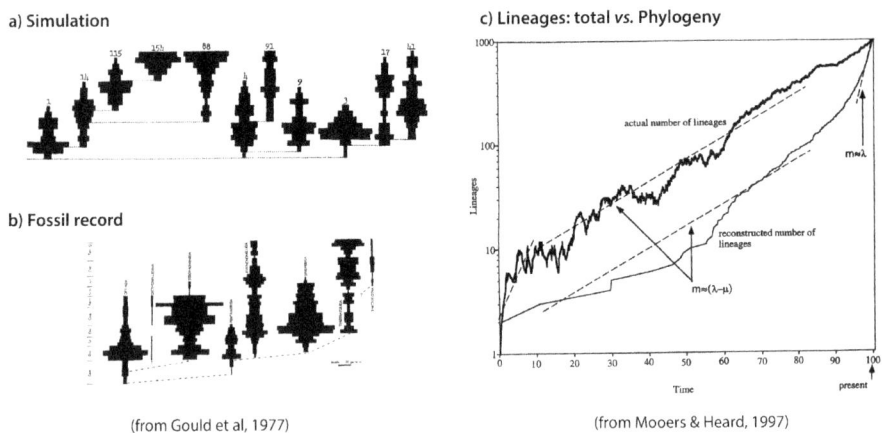

(from Gould et al, 1977)   (from Mooers & Heard, 1997)

Figure 4.3. *Lineage through Time (LTT) plot. This representation gives the number of lineages alive according to time. In the first representations, time is on the Y-axis and the width of a clade is proportional to its number of lineages. These LLT plots can be obtained a) through simulation, or b) through the analysis of fossil records. In more contemporary representations, time is in the X-axis and the number of lineages in the Y-axis (c). Owing to extinctions, the number of total lineages is always inferior to the number of lineages of the rebuilt phylogeny, except at the present time where these two numbers are equal (on the right on the LTT plot)*

The major turning point of the popularity of these birth–death models was their use to describe molecular phylogenies [NEE 94], whose branch lengths are estimated with some confidence, under the assumption of a constant mutation rate [FEL 04]. Thus, it is possible, from a phylogeny, to infer the birth and death rates of a clade with no information derived from fossils. It should be noted that these fossils are however essential to calibrate the mutational time derived from the phylogenetic inference in real time. Provided that we have an ultrametric tree – a tree where the distance between the root and all the leaves are equal – it becomes possible to calculate the likelihood of this tree under a model specified by parameters b_t and d_t [NEE 94].

The reasoning key point is to see that, in a rebuilt phylogeny, only the branches of the lineages still existing at the present time are considered. All of the lineages that became extinct in the past are invisible in the phylogeny (Figure 4.2(c)). There are always more (or as many) total lineages alive than (as) lineages in the phylogeny (those that have offspring in the present) [MOO 97]. More surprisingly, these two numbers of lineages increase at the same rate b-d, except for recent times (on the right of Figure 4.3(c)). Several lineages in the process of becoming extinct, but still alive at the present time, are included in the phylogeny and create this apparent acceleration of the diversification rate in recent times, an effect called "pull of the present".

The use of birth–death models is still very successful. We will give two examples illustrating how some model improvements made it possible to answer major biological questions on the diversification of some clades.

A first improvement is the explicit modeling of the time heterogeneity of birth and death rates. The likelihood of a birth-death model can be calculated when the rates are constant between fixed time points ("stepwise constant") [STA 11]. On either side of these points, rates change abruptly, which makes it possible to model more simply diversification rate changes over time. On a statistical basis, it is possible to determine not only the birth and death rates at the different periods separated by changes, but also the moments these rates change. This approach showed that mammals lived their period of high diversification ~30 million years ago and not at the end of the Cretaceous (~65 million years), as has been thought previously.

A second improvement involves making birth and death rates specific to some clades [MOR 11]. The rates thus estimated are then no longer a compromise on the whole phylogeny, but can reflect specific histories when some clades are declining, while others are expanding. By means of such a model, it was possible to show that cetaceans were globally declining as indicated by fossil registers, and not expanding as the previous studies based on birth–death models with constant rates indicated.

Another promising approach to analyze the diversity patterns obtained under these birth–death models is the retrospective approach. The genealogy of a clade that evolved under such a model is mathematically characterized. This coalescent approach facilitates the study of the resulting diversity for a very wide range of scenarios [LAM 13]. This retrospective vision of evolution is a recent echo of the coalescent theory developed in population genetics that is presented below.

Birth–death processes reflect a certain vision of evolution in which species are reduced to simple lineages, which are representative of the average behavior of the individuals. These processes do not make any assumption on the causes of speciation or extinction, except that they are very fast events when compared with the species life expectancy. They ignore all evolution events through composition (recombination, horizontal transfers, hybridizations), which create mosaic organisms that are only compatible with the so-called reticulated vision of evolution (description of evolution through networks). They offer an amazing thinking framework by explicitly modeling kinship between species represented in the form of trees, which provides quantifiable predictions on the expected diversity patterns. Nevertheless, we would like to advise the reader to remain somewhat lucid as to the low robustness of the inferences of ultrametric phylogenetic trees, information which is however the foundation for several methods to estimate birth and death rates. Bayesian approaches (for example, [DRU 02, LAR 09]), which sample not only the parameters of the birth–death model, but also of phylogenetic trees, seem from this point of view more robust, even though they are very costly in calculation time. Furthermore, it is crucial to measure *a posteriori* the adjustment of these models to data. For example, is the distribution of the branch lengths obtained by the adjusted model similar to that observed for the data? Even with adjusted parameters, we must consider whether the selected model correctly reproduces, at least in some respects, the data observed.

4.2. Coalescent trees

Population genetics, founded by Fisher, Wright and Haldane in the 1920s, models the evolution of the gene pool borne by the individuals of a *population*, an arbitrary grouping, which will be here assimilated with individuals of the same species. This microevolutionary approach analyzes the evolution of the different allele frequencies borne by genes over short evolutionary periods, typically in the range of a few tens of thousands generations.

Population genetics has been enriched by increasingly complex models, which are always better characterized at the conceptual level. In the 1960s, under the impulse in particular of T. Dobjansky and R. Lewontin [HUB 66], models were confronted with data, which were first proteins (migration polymorphisms according to the electric charges), and then massively genetic sequences of organisms collected in nature. This confrontation gave rise to the "neutral theory of molecular evolution" [KIM 68, KIN 69, KIM 83]. In order to explain the great diversity observed at the molecular level, this theory assumes that observed polymorphisms are almost all neutral, namely with no effect on the fitness of individuals. As directional selection tended to very quickly purge diversity, it was not considered as one of the first causes explaining intraspecific diversity.

It is within this theoretical framework that numerous quantitative predictions on molecular diversity were established (for example, [NEI 87]). These predictions were quickly again compared with data, which confirmed and thus justified the theoretical framework. Nowadays, the neutral theory is a reference model, that is, an actual null hypothesis in the statistical sense, which must be rejected in order to offer another scenario compatible with the data. Before continuing this discussion on the validity of the null model, we would like to clarify its fundamental hypotheses and show how filiations between chromosomes emerge from it.

The Wright–Fisher model (Figure 4.4(a)) helps us to understand the mechanics of genetic drift, a phenomenon that is central to the neutral theory. Individuals are reduced to a population of N chromosomes. At each generation, as many new chromosomes are created by independently copying parent chromosomes chosen *with replacement* in the previous generation. The chromosomes of the previous generation are then eliminated. In this model, some chromosomes have several offspring, while others have none. The probability that a parent chromosome has no offspring in the next generation is that of not having been chosen ($1-1/N$) by any of the new N chromosomes, namely: $(1-1/N)^N \sim e^{-1} \sim 0.38$. This genetic drift dramatically decreases diversity with each generation, replacing approximately 38% of the chromosomes of the previous generation with copies of the remaining 72%. In the long term, if no mutation is introduced, all the population chromosomes become clonal and diversity is completely lost.

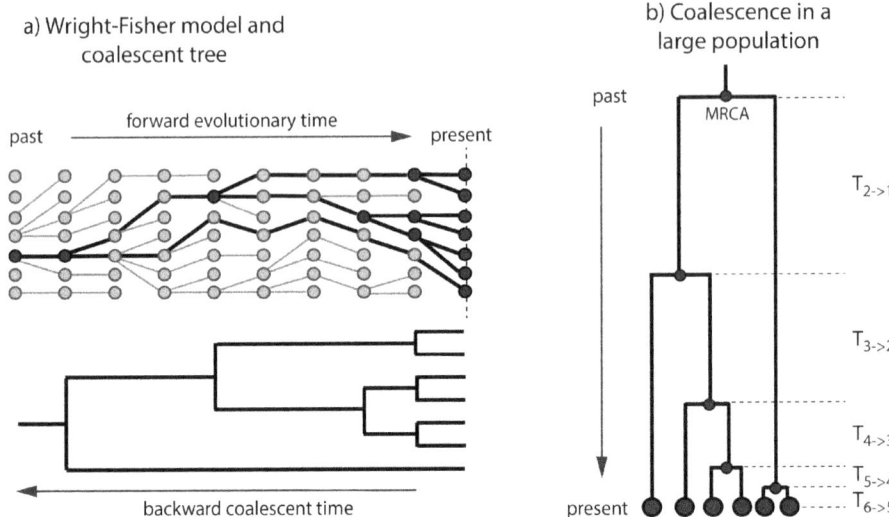

Figure 4.4. *Wright–Fisher model and coalescent tree. a) During a generation of the Wright–Fisher model, each offspring chromosome evenly draws a parent from the previous generation. This model represents a stationary equilibrium (with no beginning, nor end). The coalescent tree emerges from the backward analysis of the model, where the lineages of individuals from a sampling time are monitored. b) The coalescent tree of a population sample is characterized by successive time, in which the number of lineages alive is reduced by one, until the last common ancestor of all the sampled chromosomes is reached*

Even though this model was originally considered in forward time, its study in backward time [MAL 48] turned out to be simpler in order to predict the expected molecular diversity. The coalescent theory [KIN 82, TAJ 83, HUD 83] studies the characters of the genealogy of the individuals sampled at the present time. We will only give below a few simple principles. Readers wishing to learn more about the matter are invited to read a small introductive text in French [ACH 16] or more comprehensive books on the issue [HEI 05, WAK 09].

Let us consider the genealogy of two chromosomes randomly taken in the population. With each generation, these two chromosomes of a population have a chance $1/N$ to come from the same parent chromosome. When the population is large, this chance is small and, in most cases, both parent chromosomes are different. As long as they are disjoint, the lineages of these chromosomes then have a probability $1/N$, with each generation back in the past, of being found within the same ancestor. The coalescent time of these lineages is then a waiting time that can be approximated with a $1/N$ rate exponential law and we must go back an average of N generations to find this common ancestor.

When several chromosomes are sampled at the present time, the coalescent process is modeled by considering that each potential pair among the k lineages still alive can equally coalesce at each generation. The potential $k(k-1)/2$ pairs of these lineages correspond to the coalescent rate, which reduces the number of lineages from k to $k-1$. Thus, by sampling five chromosomes, the successive coalescent rates are $10/N$ (5 lineages -> 4 lineages), $6/N$ (4->3), $3/N$ (3->2), $1/N$ (2->1). This gradual slowing down of the coalescent rate between lineages gives the Kingman average coalescent tree small external branches and large internal branches (Figure 4.4(b)). In the Kingman original coalescent model, reductions of more than one lineage (for example, from 3 to 1 lineage) are so unlikely that they are neglected.

Unlike trees derived from the birth–death process, Kingman trees were not explicitly built by the model; they emerge from it. It is the generation construction through sampling with replacement that creates the kinship between individuals and thus builds the coalescent tree. In addition, we would like to highlight that it is only by retrospectively interpreting a sampling process, although it is prospectively defined, that the coalescent tree appears. This retrospective vision is very natural in evolution, when we talk about the genealogies of individuals or phylogenetic inferences.

The *standard neutral model* includes not only the genetic drift purging diversity, but also mutation events creating new variants: this is "mutation-drift" equilibrium. Mutations occur at a constant rate μ. Thus, with each generation, during the replication of N chromosomes which form the population, $N\mu$ new variants appear on average.

Two chromosomes with their ancestor N generations in the past are separated by $2N$ generations and therefore carry on average $2N\mu$ mutations. The average difference between two chromosomes of a population is a direct measurement of its diversity. This is due to equilibrium between mutations occurring along lineages and the genetic drift seen as the lineage coalescence. The coalescent time of these chromosomes measures their kinship.

As with trees derived from birth–death processes, these coalescent trees are compared with data. The coalescent model is remarkably simple since it is only defined by two parameters: the population size N and the mutation rate μ. A simple prediction is that the average difference between two chromosomes is $2N\mu$. Unlike the mutation rate, which has a clear biological meaning, this number N is harder to understand for an actual biological species. It is the number of chromosomes that would contribute to the next generation. This number is unknown but measurable with the average kinship between two chromosomes. This number is called the "*effective population size*" and written N_e. However, if the Wright–Fisher model does not describe well the biological reality, the meaning of this effective size becomes uncertain.

In human species, the average difference number between two haploid genomes randomly taken in the world population is approximately 1/1,300 nucleotide [SAC 01]. The average genome mutation rate is estimated at 1.2×10^8 per nucleotide per generation [KON 12]. The number of chromosomes taking part in the next generation would then be $N = \text{diff}/2\mu = (1/1,300) / (2 \times 1.2 \times 10^{-8}) \sim 32,000$ chromosomes or 16,000 diploid individuals. Yet, we are today approximately 7.5×10^9 individuals. Thus, we should be nearly 500,000 times more diverse than we actually are, if the Wright–Fisher model described our evolution well. Therefore, how can this *effective size* predicted by the standard neutral model be reconciled with the *census size* of the population? One answer at least partially satisfying to this question is population demography. The standard neutral model assumes that the number of chromosomes is constant over time. Yet, in biological species, this assumption is completely unrealistic. Even if some species oscillate around equilibrium, some are expanding, while others are declining. The human species is clearly expanding, and the population was probably small in a distant past. However, is this expansion sufficient to explain such a difference? This question is still an open field of investigation.

Another interpretation of the *effective size* seems to us more relevant to understand the evolution of the population of interest. By estimating the effective size from the average difference between two chromosomes, we in fact measure their degree of kinship. For the human species, if we accept that the generation time is identical in the different lineages, our last common ancestor "lived" on average 32,000 generations ago, or 700,000 years ago, by counting 22 years per generation. Our "average" genetic ancestor is then older than the appearance of modern humans, dated 200,000 years ago [MCD 05]. It is possible that, at that time, the number of ancestral chromosomes was very low and that they were then closely related.

Until now, we have reasoned on a single genealogy linking the chromosomes of a sample. However, the coalescent theory took into account, from the very beginning, recombination events [HUD 83] that decouple the evolutionary histories of fragments on either side of the recombination points (Figure 4.5(a)). We will call *linkage block* the chromosome area located between two recombination points. Each linkage block has its own genealogy partially connected to that of neighboring blocks. The graph describing all the genealogies of all linkage blocks is a complex mathematical object called Ancestral Recombination Graph [GRI 97] (Figure 4.5(b)), which is commonly approximated by a Markov chain [MCV 05] linking step by step the genealogies of the linkage blocks along the chromosome (Figure 4.5(c)). Thus, a recombining chromosome does not have a single genealogy, but a sequence of genealogies forming a collection of the different outputs of the coalescent process.

Thanks to the use of this very large collection of genealogies, it is possible to infer, under certain hypotheses, the demography of individuals bearing chromosomes. The number of different genealogies is so high at a genome level, that it is possible to infer the demographic history of a species from the sequence of a single diploid individual [LI 11]. By sequentially analyzing the differences between homologous chromosomes, the coalescent times of the different linkage blocks can be inferred and themselves used to infer the variation in effective sizes over time. However, this approach assumes that these variations are caused by the demography, whereas other explanations are possible, such as the structure in sub-populations [MAZ 16] or selection for some genome locations.

The place of natural selection as an explanatory process of diversity was observed to be highly regressed since the advent of the neutral theory. It is important to note the position of some authors (for example, [GIL 94]) who suggest that the standard neutral model should nevertheless be applied with a more critical eye. Its use to explain the evolution of microbes or viruses, whose population sizes are incommensurable, seems to us particularly unsuitable. The biological scope of the standard neutral model is a genuine question which we believe deserves to be largely discussed.

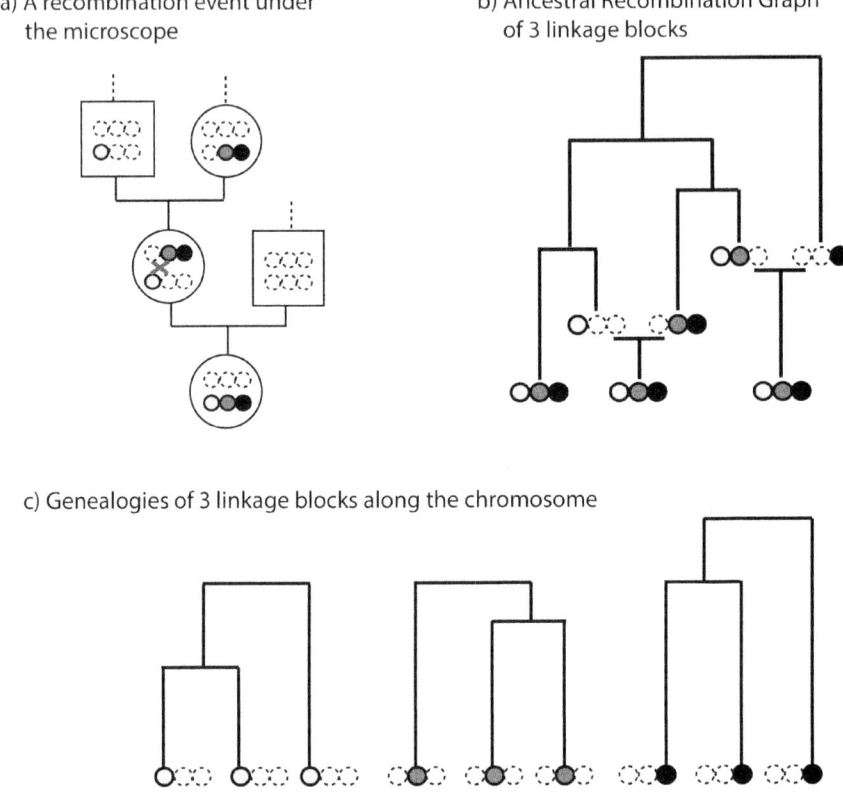

a) A recombination event under the microscope

b) Ancestral Recombination Graph of 3 linkage blocks

c) Genealogies of 3 linkage blocks along the chromosome

Figure 4.5. *Coalescence and recombination. a) Zoom on a recombination event of b) which assembles DNA fragments in prospective time (here block 1 with blocks 2 and 3) from different parents. In retrospective time, this recombination event decouples the genealogy of the linkage blocks on the right and left of the event. b) Graph representing all the genealogies of three linkage blocks. c) The sequence of the different genealogies of the ancestral recombination graph represented in b). For a color version of this figure, see www.iste.co.uk/grandcolas/biodiversity.zip*

Its relevance to model microbes or viruses is particularly questionable. The diversity of HIV-1 virus within patients is on average 1–5%. This diversity makes it possible to estimate an effective size for HIV of approximately 100, for a mutation rate of 3×10^{-5} per replication per basis [SAN 10].

The compactness of the HIV genome belies the fact that the rate of viable mutations is so high. Yet, we can independently estimate the rate of "viable" mutations and the kinship between viruses (that is, *the effective size*) from sequences collected at different points in time (Fu 2001). Applied to HIV-1, the kinship between two viruses is estimated at ~1,000 generations [ACH 04]. Yet, the viral population colonizing a patient's body that is chronically infected is at least 10^{10}! Once again, the offset between effective size and census size is colossal. As the viraemia of chronically infected patients has been stable for decades, it seems extremely unlikely that demography can explain such a difference. Another possibility is to consider that natural selection increases kinship between individuals. Each selection event results not only in the quick fixation of the beneficial mutation, but also to a lesser extent of the neighboring nucleotides, which are genetically linked to it. This effect is known as hitchhiking [SMI 74].

Sites relatively far from the selected mutation will only be partially affected by this reduction in diversity. In fact, recombination events occur between the beneficial mutation, and the neighboring sites decouple the genealogies and thus reduce the hitchhiking effects of the selected variant. In this selectionist model, the kinship between chromosomes within a species is by no means the consequence of the genetic drift, but that of the joint selection and recombination effect [GIL 00]. It is remarkable to note that, in the light of the data, it can be hard to decide between the standard neutral model and the selection recurrent event model, which both produce extremely similar diversity patterns [SCH 11, NEH 11]. However, one of the main differences is that, under the selectionist model, it is possible to observe multiple connections to the coalescent tree [TEL 14], which recall the adaptive radiations at the interspecific level.

The standard neutral model and the coalescent theory arising from it offer a stochastic interpretation of evolutionary processes, making it possible to model and figure the kinship between chromosomes within a species. They can be used to make quantitative predictions about the expected diversity between the individuals within a species. Conversely, it is possible to infer the parameters of this model, including of the demography or structuring into sub-populations, from data. Unfortunately, the standard neutral model is sometimes indiscriminately used, with no critical eye. The case of the competitive selectionist model, which produces nearly identical diversity patterns, is enlightening: even when a model correctly predicts part of the data, nothing guarantees its biological reality. Only the accumulation of independent biological observations and a deeper reflection on the meaning of the models can help us to offer a substantiated evolutionary scenario compatible with our observations. No model is innocent.

4.3. Birth–death and/or coalescent model?

We briefly presented above two classes of models offering two complementary visions of the processes creating kinship between individuals or between species. Although different, these models provide a framework with which to understand the origin of kinship phylogenetic trees. We would like to focus on the difference between phylogenetic reconstruction in the standard sense and the models we described above. The subject of phylogenetic inference is the reconstruction of an "actual", or at least credible, tree from a set of character data [FEL 04]. This approach is part of the characterization of a biodiversity *pattern*. The models described above provide *a contrario* an assumed theoretical framework in which *processes* generate trees from evolutionary mechanisms.

Both kinship levels (individuals and species) are not independent: a species is composed of a population of individuals bearing chromosomes. The processes creating gene trees are then nested in those creating species trees (Figure 4.6). Gene genealogies can then be modeled as coalescence conditioned by a species tree. Yet, chromosomes contain a very large number of coalescent process outputs constrained by the species tree [MAD 97]. It is then possible to use these multiple outputs to infer the species tree. In fact, no coalescence can be seen between two different species, as long as the sampled chromosome lineages do not go back in the ancestral species. When the number of sampled chromosomes and the number of species are not too high, it is possible to estimate both the species tree and the population parameters explaining the chromosome trees [YAN 10].

superior organisation level

inferior organisation level

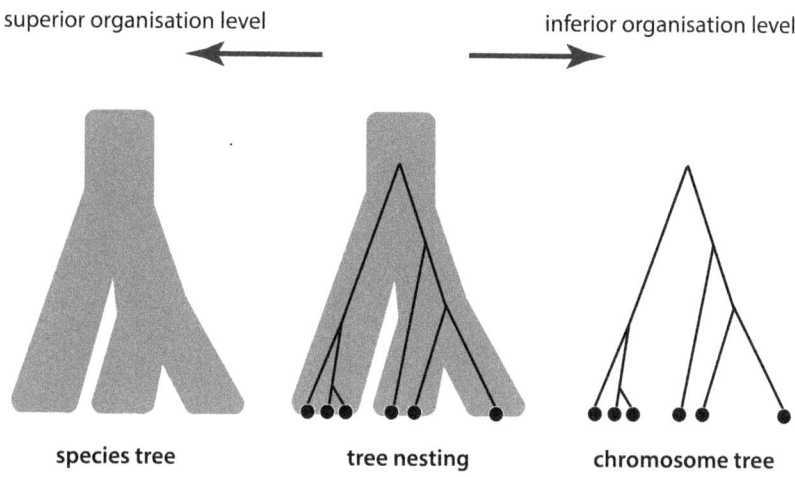

species tree   tree nesting   chromosome tree

Figure 4.6. *Tree nesting at different levels. Although evolutionary processes simultaneously occur at all the levels, it generates both species trees and chromosome trees (they are linking block trees, to be more specific)*

It would be tempting to continue this *mise en abyme* towards inferior levels, where the genealogies of repeated elements (transposons, duplicated genes, etc.) occur within chromosomes. This nesting is also continued towards higher levels, structuring diversity into families or taxonomic orders. From this *mise en abyme*, the impression arises that we are still only at the beginning of our formalization of kinship between biological entities. What are the relationships between these different levels of tree nesting? Using chromosome trees to infer the species tree is a start, but the species tree is in this case considered as a simple output to be estimated. It is a phylogenetic inference. Process interweaving would deserve further reflection.

Out of historical interest, let us note that there is a genealogy of genealogy models. David G Kendall, the man who formalized birth–death models, supervised part of John F.C. Kingman's thesis, the father of the coalescent theory. Thus, our *mise en abyme* of processes also continues at the historical level.

The dichotomy that associates the birth–death process with model interspecific diversity and the coalescent process with intraspecific diversity is surprising. Although they have an obvious historical explanation, the biological reasons for such a choice still need to be discussed. In fact, it is possible to model interspecific diversity through a coalescent approach assuming that the total number of lineages is constant over time [HEY 92, MOR 10, HUB 01], and even oscillates around equilibrium [RAU 73]. Arguments against this modeling for macroevolution are not well substantiated and are reminiscent of the selectionist–neutralist debate, which animated population genetics in the 1970s [NEI 05]. Conversely, the use of birth–death models in population genetics is nearly non-existent, although some attempts were successful [LAP 17].

Both models are never equivalent, since the standard neutral model is an equilibrium model whereas the birth–death model is never balanced. It inevitably results in the complete extinction of all lineages (*critical* and *subcritical* regimes) or the exponential growth of lineages (*supercritical* regime). We believe that we must now discuss the systematic comparison of the diversity patterns obtained by these two classes of models and understand in what way they are similar and in what way they are different. The unification of these two approaches would be a giant leap in our understanding of evolutionary processes and therefore our capacity to understand what shapes biodiversity at all levels.

4.4. Acknowledgements

I would like to thank my colleague and friend Amaury Lambert, as an endless source of scientific wonder, who generously helped me with this chapter, as well as Sophie Brouillet for her keen eye and benevolence. I would also like to thank Marie-Christine Maurel for having arranged this book. Thanks to her, I was able to write these few ideas and clear my head regarding evolutionary models.

4.5. Bibliography

[ACH 04] ACHAZ G., PALMER S., KEARNEY M. *et al.*, "A robust measure of HIV-1 population turnover within chronically infected individuals", *Molecular Biology and Evolution*, vol. 21, no. 10, pp. 1902–1912, doi:10.1093/molbev/msh196, 2004.

[ACH 16] ACHAZ G., "Introduction À La Coalescence", available at: http://wwwabi. snv.jussieu.fr/achaz/coalescence.pdf, 2016.

[BEE 15] BEERENWINKEL N., SCHWARZ R.F., GERSTUNG M. *et al.*, "Cancer evolution: mathematical models and computational inference", *Systematic Biology*, vol. 64, vol. 1, p. e1–25, doi:10.1093/sysbio/syu081, 2015.

[DRU 02] DRUMMOND A.J., NICHOLLS G.K., RODRIGO A.G. *et al.*, "Estimating mutation parameters, population history and genealogy simultaneously from temporally spaced sequence data", *Genetics 161*, vol. 161, no. 3, pp. 1307–1320, 2002.

[FEL 04] FELSENSTEIN J., *Inferring Phylogenies*, Sinauer Associates, Sunderland, 2004.

[FRA 02] FRANK S.A., *Immunology and Evolution of Infectious Disease*, Princeton University Press, 2002.

[FU 01] FU Y.X., "Estimating mutation rate and generation time from longitudinal samples of DNA sequences", *Molecular Biology and Evolution*, vol. 18, no. 4, pp. 620–26, 2001.

[GAL 75] GALTON F., WATSON H.W., "On the probability of the extinction of families", *Journal of the Royal Anthropological Institute*, vol. 4, pp. 138–44, 1875.

[GIL 94] GILLESPIE J.H., *The Causes of Molecular Evolution*, Oxford University Press, available at: http://public.eblib.com/choice/publicfullrecord.aspx?p=430880, 1994.

[GIL 00] GILLESPIE J.H., "Genetic drift in an infinite population. The pseudohitchhiking model", *Genetics*, vol. 155, no. 2, pp. 909–919, 2000.

[GOU 77] GOULD S.J., DAVID M., RAUP T.J. *et al.*, "The shape of evolution: A comparison of real and random clades", *Paleobiology*, vol. 3, pp. 23–40, 1977.

[GRI 97] GRIFFITHS R.C., MARJORAM P., "An ancestral recombination graph", *Progress in Population Genetics and Human Evolution*, in DONNELLY P., TAVARÉ S. (eds), vol. 87, pp. 257–70, Springer New York, available at: http://link.springer.com/ 10.1007/978-1-4757-2609-1_16, 1997.

[HAR 97] HARTL D.L., CLARK A.G., *Principles of Population Genetics*, 3rd ed. Sinauer Associates, Sunderland, 1997.

[HEN 05] HEIN J., SCHIERUP M.H., WIUF C., *Gene Genealogies, Variation and Evolution: A Primer in Coalescent Theory*, Oxford University Press, 2005.

[HEN 11] HENG L., DURBIN R., "Inference of Human Population History from individual whole-genome sequences", *Nature*, vol. 475, no. 7357, pp. 493–96, doi:10.1038/nature10231, 2011.

[HEY 92] HEY J., "Using Phylogenetic Trees to Study Speciation and Extinction", *Evolution*, vol. 46, no. 3, pp. 627–640, 1992.

[HUB 66] HUBBY J.L., LEWONTIN R.C., "A molecular approach to the study of genic heterozygosity in natural populations. I. The number of alleles at different loci in drosophila pseudoobscura", *Genetics*, vol. 54, no. 2, pp. 577–94, 1966.

[HUB 01] HUBBELL S.P., *The Unified Neutral Theory of Biodiversity and Biogeography*, Princeton University Press, 2001.

[HUD 83] HUDSON R.R., "Properties of a neutral allele model with intragenic recombination", *Theoretical Population Biology*, vol. 23, no. 2, pp. 183–201, doi:10.1016/0040-5809(83) 90013-8, 1983.

[JOT 05] JOTUN H., SCHIERUP M.H., WIUF C., *Gene Genealogies, Variation and Evolution: A Primer in Coalescent Theory*, Oxford University Press, 2005.

[KEN 48] KENDALL D.G., "On the generalized 'Birth-and-Death' process", *The Annals of Mathematical Statistics*, vol. 19, no. 1, pp. 1–15. oi:10.1214/aoms/ 1177730285, 1948.

[KIM 68] KIMURA M., "Evolutionary rate at the molecular level", *Nature*, vol. 217, no. 5129, pp. 624–26, doi:10.1038/217624a0, 1968.

[KIM 83] KIMURA M., *The Neutral Theory of Molecular Evolution*, Cambridge University Press, 1983.

[KIN 69] KING J.L., JUKES T.H., "Non-darwinian evolution", *Science*, vol. 164, no. 3881, pp. 788–98, New York, 1969.

[KIN 82] KINGMAN J.F.C., "The Coalescent", *Stochastic Processes and Their Applications*, vol. 13, no. 3, pp. 235–48, doi:10.1016/0304-4149(82)90011-4, 1982.

[KON 12] KONG A., FRIGGE M.L., MASSON G. *et al.*, "Rate of de novo mutations and the importance of father's age to disease risk", *Nature*, vol. 488, no. 7412, pp. 471–475, doi:10.1038/nature11396, 2012.

[LAM 13] LAMBERT A., STADLER T., "Birth–death models and coalescent point processes: the shape and probability of reconstructed phylogenies", *Theoretical Population Biology*, vol. 90, pp. 113–28, doi:10.1016/j.tpb.2013.10.002, December 2013.

[LAP 17] LAPIERRE M., LAMBERT A., ACHAZ G., "Accuracy of demographic inferences from the site frequency spectrum: The case of the yoruba population", *Genetics*, vol. 206, no. 1, pp. 439–49, doi:10.1534/genetics.116.192708, 2017.

[LAR 09] LARTILLOT N., LEPAGE T., BLANQUART S., "PhyloBayes 3: A bayesian software package for phylogenetic reconstruction and molecular dating", *Bioinformatics*, vol. 25, no. 17, pp. 2286–88, doi:10.1093/bioinformatics/btp368, 2009.

[LI 11] LI H., DURBIN R., "Inference of human population history from individual whole-genome sequences", *Nature*, vol. 475, no. 7357, pp. 493–496, July 2011.

[MAD 97] MADDISON W.P., "Gene trees in species trees", *Systematic Biology*, vol. 46, no. 3, pp. 523–36, doi:10.1093/sysbio/46.3.523, 1997.

[MAL 48] MALÉCOT G., *Les Mathématiques de L'hérédité*, Masson & Cie, Paris, 1948.

[MAZ 16] MAZET O., RODRÍGUEZ W., GRUSEA S. *et al.*, "On the importance of being structured: instantaneous coalescence rates and human evolution – lessons for ancestral population size inference?", *Heredity*, vol. 116, no. 4, pp. 362–71, doi:10.1038/hdy.2015.104, 2016.

[MCD 05] MCDOUGALL I., BROWN F.H., FLEAGLE J.G., "Stratigraphic placement and age of modern humans from kibish, Ethiopia", *Nature*, vol. 433, no. 7027, pp. 733–736, doi:10.1038/nature03258, 2005.

[MCV 05] MCVEAN G.A.T., CARDIN N.J., "Approximating the coalescent with recombination", *Philosophical Transactions of the Royal Society B: Biological Sciences*, vol. 360, no. 1459, pp. 1387–1393, doi:10.1098/rstb.2005.1673, 2005.

[MOO 97] MOOERS A.O., HEARD S.B., "Inferring evolutionary process from phylogenetic tree shape", *The Quarterly Review of Biology*, vol. 72, no. 1, pp. 31 54, doi:10.1086/419657, 1997.

[MOR 10] MORLON H., POTTS M.D., PLOTKIN J.B., "Inferring the dynamics of diversification: a coalescent approach", in HARVEY P.H. (ed.), *PLoS Biology*, vol. 8, no. 9, p. e1000493, doi:10.1371/journal.pbio.1000493, 2010.

[MOR 11] MORLON H., PARSONS T.L., PLOTKIN J.B., "Reconciling molecular phylogenies with the fossil record", *Proceedings of the National Academy of Sciences*, vol. 108, no. 39, pp. 16327–16332, doi:10.1073/pnas.1102543108, 2011.

[MOR 14] MORLON H., "Phylogenetic approaches for studying diversification", in MOOERS A.O. (ed.), *Ecology Letters,* vol. 17, no. 4, pp. 508–25, doi:10.1111/ele.12251, 2014.

[NEE 94a] NEE S., HOLMES E.C., MAY R.M. *et al.*, "Extinction rates can be estimated from molecular phylogenies", *Philosophical Transactions of the Royal Society of London. Series B, Biological Sciences*, vol. 344, no. 1307, pp. 77–82, doi:10.1098/rstb.1994.0054, 1994.

[NEE 94b] NEE S., MAY R.M., HARVEY P.H., "The reconstructed evolutionary process", *Philosophical Transactions of the Royal Society B: Biological Sciences*, vol. 344, no. 1309, pp. 305–11, doi:10.1098/rstb.1994.0068, 1994.

[NEE 06] NEE S., "Birth–death models in macroevolution", *Annual Review of Ecology, Evolution, and Systematics*, vol. 37, no. 1, pp. 1–17, doi:10.1146/annurev.ecolsys.37.091305.110035, 2006.

[NEH 11] NEHER R.A., SHRAIMAN B.I., "Genetic draft and quasi-neutrality in large facultatively sexual populations", *Genetics*, vol. 188, no. 4, pp. 975–96, doi:10.1534/genetics.111.128876, 2011.

[NEI 87] NEI M., *Molecular Evolutionary Genetics*, Columbia University Press, New York, 1987.

[NEI 05] NEI M., "Selectionism and neutralism in molecular evolution", *Molecular Biology and Evolution*, vol. 22, no. 12, pp. 2318–2342, doi:10.1093/molbev/msi242, 2005.

[RAU 73] RAUP D.M., GOULD S.J., SCHOPF T.J.M. *et al.*, "Stochastic models of phylogeny and the evolution of diversity", *The Journal of Geology*, vol. 81, no. 5, pp. 525–42, 1973.

[SAC 01] SACHIDANANDAM R., WEISSMAN D., SCHMIDT S.C. *et al.*, "A map of human genome sequence variation containing 1.42 Million single nucleotide polymorphisms", *Nature*, vol. 409, no. 6822, pp. 928–933, doi:10.1038/35057149, 2001.

[SAN 10] SANJUAN R., NEBOT M.R., CHIRICO N. *et al.*, "Viral mutation rates", *Journal of Virology*, vol. 84, no. 19, pp. 9733–9748, doi:10.1128/JVI.00694-10, 2010.

[SCH 11] SCHIFFELS S., SZOLLOSI G.J., MUSTONEN V. *et al.*, "Emergent neutrality in adaptive asexual evolution", *Genetics*, vol. 189, no. 4, pp. 1361–75, doi:10.1534/genetics.111.132027, 2011.

[SMI 74] SMITH J.M., HAIGH J., "The hitch-hiking effect of a favourable gene", *Genetical Research*, vol. 23, no. 1, pp. 23–35, 1974.

[STA 81] STANLEY S.M., SIGNOR P.W., LIDGARD S. *et al.*, "Natural clades differ from 'random' clades: simulations and analyses", *Paleobiology*, vol. 7, no. 01, pp. 115–127, doi:10.1017/S0094837300003833, 1981.

[STA 11] STADLER T., "Mammalian phylogeny reveals recent diversification rate shifts", *Proceedings of the National Academy of Sciences*, vol. 108, no. 15, pp. 6187–6192, doi:10.1073/pnas.1016876108, 2011.

[TAJ 83] TAJIMA F., "Evolutionary relationship of DNA sequences in finite populations", *Genetics*, vol. 105, no. 2, pp. 437–460, 1983.

[TEL 14] TELLIER A., LEMAIRE C., "Coalescence 2.0: a multiple branching of recent theoretical developments and their applications", *Molecular Ecology*, vol. 23, no. 11, pp. 2637–2652, doi:10.1111/mec.12755, 2014.

[WAK 09] WAKELEY J., *Coalescent Theory: An Introduction*, Roberts & Co. Publishers, Greenwood Village, 2009.

[YAN 10] YANG Z., RANNALA B., "Bayesian species delimitation using multilocus sequence data", *Proceedings of the National Academy of Sciences*, vol. 107, no. 20, pp. 9264–9269, doi:10.1073/pnas.0913022107, 2010.

[YUL 24] YULE G.U., "A mathematical theory of evolution, based on the conclusions of Dr. J.C. Willis, F.R.S", *Philosophical Transactions of the Royal Society B: Biological Sciences*, vol. 213, nos 402–410, pp. 21–87, doi:10.1098/rstb.1925.0002, 1924.

5

Analysis of Microbial Diversity: Regarding the (Paradoxical) Difficulty of Seeing Big in Metagenomics

5.1. Introduction

The wealth of biological communities is extraordinary. The complexity of these communities' organization as well as the diversity of their (gene, function, organism, taxon) composition are major challenges for scientists developing methods to produce as realistic descriptions and recognition of the living world as possible. To understand the causes of biodiversity, it seems in particular essential to be able to see big, i.e. to be able to list the main actors (for want of all actors), the main relationships between them (for want of all their relationships) and the main processes (for want of all processes) structuring biological communities, and then compare these organizations with each other in space and time. Can this ambitious objective, with a holistic inspiration, be achieved? In fact, molecular data have never been so abundant or so easy to obtain as nowadays. As for microbial communities, metagenomics provides a colossal wealth of data: gene, RNA and even genome sequences of prokaryotes; protists and their mobile elements (viruses, plasmids) present in an environment. For example, the Human Microbiome Project (HMP) implemented 1,265 metagenomics projects between 2011 and 2014 and thus produced more than 80 million gene sequences (see http://hmpdacc.org/catalog/grid.php?dataset= metagenomic). The Tara Projects studied 210 oceanic ecosystems from

Chapter written by Chloé Vigliotti, Philippe Lopez and Eric Bapteste.

20 biogeographic areas [PES 15] and provided more than 111 million gene sequences [SUN 15]. These emblematic projects are far from being the only ones; the EBI thus lists 67,375 samples from 1,073 metagenomics projects (see https://www.ebi.ac.uk/metagenomics/search).

These data are obtained through several stages. First, there is a sampling phase (water, soil, feces, etc. are collected). Then, these samples are filtered, in particular by size fraction, which makes it possible to separate microorganisms of different sizes from viruses, etc. Then, the DNA is isolated, amplified and sequenced. Finally, molecular data are analyzed by bioinformatic pipelines.

Through its approach, metagenomics is then part of an emerging mode of science: data gathering by means of methods already developed outside the laboratories using them [KRO 12]. For example, the first stages of the lizard gut microbiome study, after the lizards are captured and the DNA contained in their stomach is isolated, consist in choosing a sequencing method and identifying a service provider that will implement it after receiving the samples. Science philosopher U. Krohs described this type of data production procedure, which is massive, simple and relatively pre-prepared, as "convenience science" or "automated science" [KRO 12]. As sequencing is affordable (obtaining 104 million read pairs per 2×300 bp illumina costs $17,500 in 2015), most of the laboratories in the world collect their data that way, and the amount of metagenomic databases publicly available, such as MGRAST (http://metagenomics.anl.gov), is exploding. We could then naively assume that metagenomics provides a very wide view of microbial communities. In other words, we might hope that it would be possible to produce generalizations, models and even theories on the composition, organization, dynamics and evolution of microbial communities, by comparing all these data, even though at present no database centralizes most of the sequences obtained. However, replying in the affirmative would come to reduce a paradox at the heart of these readily inclusive approaches: their view extent and depth are restricted by epistemic factors and practices, which we will present in this chapter by relying mainly on our experience in the analysis of microbiomes (i.e. microbial community genes) and lizard gut microbiota (i.e. microbial community taxa) from the *Podarcis sicula* species, which is characterized by diet variations. In fact, over 35 years, a population of these lizards, which were originally insectivore, became omnivore by integrating into their diet nearly 80% of plants [HER 08].

5.2. Comparing metagenomic data sets is difficult

Metagenomic data sets (or microbiomes) are both numerous and very diverse, which makes their comparison difficult. This diversity of data sets is found at several levels. First, environmental metagenomes must be differentiated (for example, TARA oceans [PES 15, SUN 15, VAR 15]) from those associated with hosts. With regard to host-associated metagenomes in particular, there are a variety of studies on microbiomes associated with a wide variety of hosts: microbiomes [WAL 11] associated with humans [HUM 12], mice [XU 16, ZHE 16], gorillas [GOM 15, MOE 15], chimpanzees [MOE 12], frogs [KOH 13], pandas [WEI 15, ZHU 11], termites [SU 16], cows [MCC 14, YAN 15], iguanas [HON 11], rabbits [ZEN 15], bees [MAR 11], etc. The association of metagenomes with hosts generates here again a data set diversity depending on the sequenced microbiome [HUM 12]: oral [SCH 17], skin [PRA 17], intestinal [WAL 11, LEC 13, TUR 09a, YAT 12, GIL 06, TUR 09b], vaginal [MA 12, MED 17], etc. It could even be argued that a lot of these microbiomes are partly immeasurable. This is mainly due to the sequence acquisition modes that are very heterogeneous. Even within "convenience science", there are a variety of sequencing methods. Among the main sequencing technologies used are SOLiD, Ion Torrent PGM (Life Sciences), HiSeq 2000, MiSeq (Illumina) and 454 (Roche) [LIU 12]. These different technologies have different characteristics, particularly regarding the read length produced (for example, in 2012, 454 GS LFX produced 700 base pair reads, while HiSeq 2000 produced 50–100 base pair reads), in terms of accuracy (99.9% for 454 compared to 98% for HiSeq), time and monetary cost [LIU 12] and sequencing biases (specific to each technology), which create data sets that are difficult to compare to each other. In fact, this is reflected by the constitution of data sets with variable sequence quantities, qualities and lengths, which actually capture different proportions of environmental DNA (which is measured by the sequencing cover of microbiomes and saturation curves) and are then unequally representative of these communities. These experimental biases limit microbiome comparisons (as well as those of microbiota), even for close systems. The effect of the sequencing year was, for example, particularly visible in our studies. We obtained approximately 1 million V4 region sequences for a DNA marker (16S rRNA) in 2014 and 1.7 million sequences of the V4 region in 2015, in order to characterize the composition of microbial communities in the cecal valve of *P. sicula*. In the first year, these had been produced by an initial sequencing method (2 × 300 base pairs by Illumina); in the second year, a different method had been used

(2 × 250 base pairs by Illumina). The taxonomic annotation of these sequences made it possible to differentiate two main types of microbial communities within these lizards. After verification, we were able to conclude that this fascinating result did not reflect an essential aspect of lizard biology, but a sequencing bias. Lizard microbiomes sequenced using the same method looked more like each other than like the microbiomes of other lizards (see Figure 5.1).

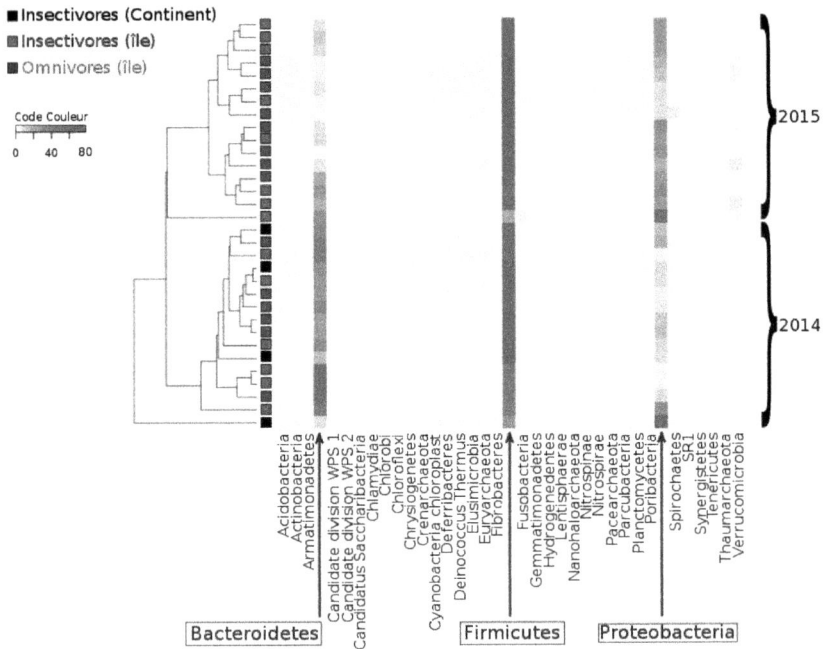

Figure 5.1. *Impact of the sequencing method on the taxonomic composition of lizard microbiota. For a color version of this figure, see www.iste.co.uk/grandcolas/biodiversity.zip*

Another significant bias is that the number of sequenced individuals can have an impact on the results obtained. The analyses of human enterotypes (i.e. singular microbial communities with a close taxonomic composition) thus revealed different results according to the number of sampled individuals. The study of 39 samples from different European countries, America and Japan [ARU 11] concluded the existence of three enterotypes, whereas a subsequent study conducted on 98 samples provided only two enterotypes [WU 11]. This heterogeneity complicates the interpretation of comparative metagenomic analyses, because it would require being able to differentiate the signals reflecting sequencing method variations and those reflecting the biological properties of microbiomes. In other words, the abundance of data sets does not immediately provide metageneticists with the elements that would allow them to perform syntheses easily.

In addition, the nature of the sequenced molecules significantly reduces the scope of biodiversity on which metagenomics focuses. A large number of data sets in fact target specific regions of a specific marker, 16S rRNA (for prokaryotes) and 18S rRNA (for eukaryotes). The choice of these regions and molecules makes it possible to build primers (based on known sequences) in order to identify their homologs in environments. The choice of this marker is explained by the fact that researchers would like to be able to associate 16S (or 18S) forms diverging by more than 3% between them, with distinct species. In this context, collecting these molecules can in particular be used to infer, as a first approximation, microbiota, i.e. the taxa composition of the microbiome. However, this approach increasingly substantially ignores numerous organisms. Thus, CPR bacteria, which are ultra-small and with a fascinating biology, although they are ubiquitous in the environment, had not been discovered because their 16S was atypical [BRO 15]. This approach also ignores the massive viral diversity; several rare microbes are poorly detected, only because of their low individual abundance, even if they collectively represent significant portions of communities. In fact, within a microbial community, a minority of abundantly present species and a large majority of rare species coexist [JOU 17, NEM 11]. These rare microbes form what is called the "rare biosphere". More specifically, the rare biosphere corresponds to all microbial species abundant in the ecosystem considered, not exceeding 0.1% of the total microbial community [WAN 17]. If one liter of water is considered, the rare biosphere present in this ecosystem includes hundreds, or even thousands, of species of bacteria and archaea [WAN 17]. Metageneticists then know that they are not only analyzing all the sequences of an environment but also assigning all of them to known taxa. Non-annotated sequences (at the taxonomic level as well as the functional level) even form a microbial dark matter [LOP 15]: numerous actors in microbial communities with a role and history that are difficult to infer. In addition, we will see later that knowing the diversity of 16S molecules does not mean knowing the diversity of the other genes present in microorganisms.

Consequently, there is a paradox: metagenomics offers a view that is both wide and narrow. On the one hand, sequences are more abundant than ever and, on the other hand, they are not necessarily easily comparable, nor necessarily representative. It is therefore crucial to identify means to widen the view of microbial diversity, which could *prima facie* be first performed by following two very distinct research branches. We will present these branches and check whether they can achieve the stated objective: enriching the knowledge on microbiomes by providing a wider view. However, to do so, a prior detour and reflection are necessary in order to check whether metagenomics, such as it is practiced nowadays, has taken a path dependency or not.

5.3. Path dependency and knowledge production

Several science philosophers have wondered to what extent data-producing practices and technologies affect scientific knowledge production [LEO 12]. They agree on the fact that data growth and the developments of technologies to obtain and process them lead to conceptual changes in sciences and the way scientists assess, compare, interpret and reuse the data sets available. Thus, U. Krohs [KRO 12], by relying in particular on the great work of S. Sydow on path dependency [SYD 05], suggested that, like the development of social organizations, knowledge production works in stages, and that "convenience science" tends to lead scientists to conformist path dependency. Nonetheless, to apply this conclusion to system biology, and in order to clarify the situation of metagenomics, we must present in a few words what we mean by the notion of epistemic path dependency.

We will rely to do so on the famous diagram by S. Sydow [SYD 05] (see Figure 5.2), which we will reinterpret in the context of microbial community study. According to this diagram, the first stage of knowledge production (phase I) is characterized by a profusion of questions, resulting from a diversity of data analysis and production methods. The heterogeneously obtained knowledge is therefore little, or not at all, comparable, and *explananda* flourish, forming a wide catalog of observations. The methods used during this first phase are neither completely arbitrary nor random: they depend on the history of the disciplines that preceded metagenomics, but they are very little constrained and only poorly selected. Progression to a second phase (phase II) occurs when the choices of analysis methods start to converge, in particular because the use of certain approaches (rather than others) leads to rewards. Beyond this critical point, a positive feedback sorts through methods and therefore through the *explananda* that will be the study subject of the discipline. This is understood very intuitively: if some methods generate publications resulting in funding, then these methods can be given priority in future studies by a snowball effect. When the methods of several scientists thus start to converge, generalizations in respect of the phenomena studied become in principle possible, because comparable work can be compared to each other. Nevertheless, during this second phase, all the knowledge is not exclusively produced by a limited number of approaches. Several works remain practically and conceptually heterogeneous. The discipline is therefore not yet on a path dependency: it is simply starting to find its path. The situation can nonetheless move to a third stage (phase III) when a locking occurs: a single or a few approaches are required from all newcomers. "Convenience science" encourages this locking. It is the famous "biology kitification", because many experimental procedures are standardized and based on kits provided by the industry. This third phase then facilitates the production of standardized *explananda*, which makes it possible to increase knowledge by allowing maximum comparisons between studies, but paradoxically limits their scope because they all focus on the same specific aspect of biological diversity. This even actually results in the promotion of a specific ontology of such studies, in the image of the use of Operational Taxonomic Units (OTUs). An OTU [SNE 62] is defined as the base unit facilitating the grouping of phylogenetically close individuals. Within a sample, the construction of OTUs then requires the scientist to compare all the sample sequences in pairs and to consider that two sequences belong to the same OTU if the similarity between these sequences is superior to a selected threshold value. The selected threshold value, associated with 16S rRNA, is often 97% of similarity between two sequences, because it is commonly accepted that an OTU defined at 97% of similarity corresponds to one species [KON 05]. "Biology kitification" is reflected here by the use of QIIME [CAP 10, SHA 14] in numerous studies to construct OTUs, which will then be used by this software to carry out a series of analyses (taxonomic annotations, diversity analyses, etc.). This choice of methods will give rise to theories. Nevertheless, the types of results obtained in phase III are also characterized by blind spots, invisible *explananda*. Such standardization is potentially liable to prevent scientists from making even more interesting discoveries. This is why S. Sydow [SYD 05] emphasizes the possibility of a last phase of knowledge production (phase IV): the exit from a path dependency. This approach must be particularly active to succeed in unlocking the characteristic practices of a discipline, even when these practices do not achieve optimal results. By analogy, it is very difficult to change computer keyboards, even though the organization of the keys is no longer justified in the same way as when QWERTY keyboards were invented with typewriters.

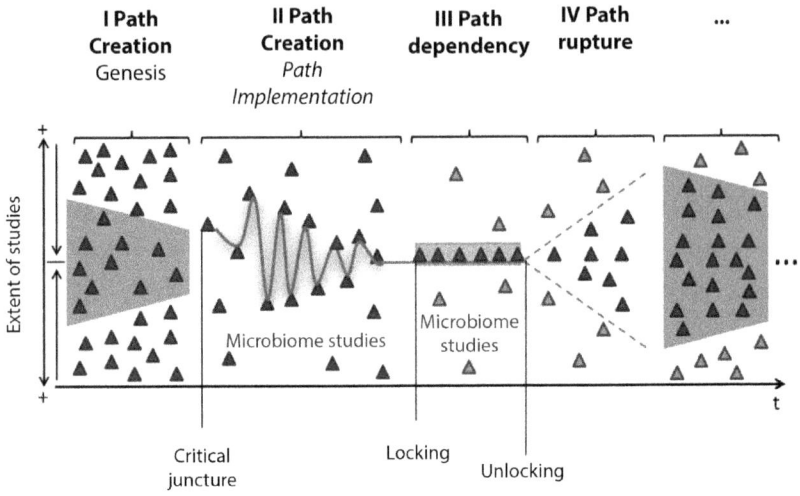

Figure 5.2. *Hypothetical sequence of stages making it possible to acquire knowledge in a scientific discipline, adapted from Sydow et al. [SYD 05]. For a color version of this figure, see www.iste.co.uk/grandcolas/biodiversity.zip*

In relation to these phases, where do studies of the microbial community diversity by metagenomics lie? These seem to us to be in a hybrid situation, which is different depending on whether they are microbiome studies or microbiota studies. Undoubtedly, beyond sampling (which can have many points in common between different studies), data acquisition uses several parallel approaches, which are differentiated by technical details, but which conceptually aspire to produce the same type of molecular data. There is then a corridor of sequencing methods that are relatively similar rather than different, which especially results in the production of numerous 16S and 18S reads, in the case of metagenomic analyses focused on a marker, and in the production of numerous reads from random genome regions, in the case of non-targeted metagenomics. The difference between these sequences mainly concerns the read length and the type of sequencing errors associated with each method. This low diversity of data production methods is perfectly consistent with the existence of a "convenience science", because the aim of sequencing method developers is to transform as fast as possible the status of these research technologies from new to standard.

However, the next analysis phase ("data mining") to understand these metagenomics data is for now less coercive, even if the great questions to be answered are already asked. More specifically, in respect of genes as taxa, metageneticists mainly wonder "who is there?", "who does what?" and "which communities are the most diverse?" In other words, there is a high selection of methods, not because of their implementation, which can be different, but according to the questions they will help to answer, which are generally the same. These observations place metagenomics at least in phase II of the Sydow diagram, i.e. close to a path dependency. In fact, we think that, because of the diversity of bioinformatic pipelines implemented during the "data mining" stage, microbiome studies can probably be assigned to this phase II. In fact, there are several ways to answer the question "what are the taxa present in the microbiome studied?" On the one hand, the study can focus on some specific genes detected in the microbiome. These phylogenetic markers are thus compared to their reference database homologs (homology searching), in order to receive a taxonomic assignment, and then be included in a phylogenetic tree [SHA 14]. This method is based on classification algorithms and the similarity of reads compared to gene marker sequences. Several tools can be used to answer the question asked with this method: MetaPhyler [LIU 11], MetaPhlAn [SEG 12], AMPHORA [WU 08, WU 12], etc. On the other hand, various approaches, called binning approaches, can be used to differentiate the different taxa present in a metagenome [SHA 14]. Microbiome reads can also be assembled into contigs (independently or following binning). This concatenation of small DNA fractions with size varying between 35 base pairs and several hundred base pairs (according to the sequencing method) facilitates the taxonomic and functional annotation of the microbiome content, while providing information on the genomic context in which microbial genes are found. There are several tools to perform these assemblies: IDBA [PEN 12], MetaVelvet [NAM 12], metaSPAdes (spades.bioinf.spbau.ru/release3.10.0/manual.html), Ray Meta [BOI 12], MetAMOS [TRE 13, TRE 11], etc. In addition, all microbiome analyses do not necessarily relate to the same questions. While some wonder which taxa are present in the community, others wonder instead: which genes are present in the community? How do microbes interact within the community? What is the genetic diversity of a microbiome? In which genomic contexts is a given gene found? In short, microbiome studies involve different methods, each using several tools, the use of which requires the selection of a significant number of parameters. This succession of methods and heuristics submerges the analyst with numerous dilemmas.

Nevertheless, for microbiota studies (probably because the original data are even more homogeneous), pipelines and "good practices" are clearly more defined and limited. Microbial diversity analyses (based on 16S or 18S) seem to have entered phase III, i.e. to have taken a path dependency. If we take the example of the question "Who is there?" for a microbiota study, whose data to be analyzed are 16S rRNAs, there is a path dependency commonly taken: first, OTUs are defined at 97% of identity. Then, these OTUs are taxonomically assigned by comparing OTU centroid sequences to a reference database. These taxonomic assignations then make it possible to calculate diversity within each sample (alpha diversity) and then to compare sample diversity in pairs (beta diversity). Finally, taxonomic assignations are used to calculate the relative abundances of each taxon per sample. To our knowledge, if this way to answer the question "Who is there?" is standard, there is still however a diversity of tools and measurements for each stage: the QIIME tool is the most commonly used to construct OTUs, but it uses several algorithms to perform this task (pick_de_novo_otus.py, pick_closed_reference_otus.py, pick_open_reference_otus.py). There are however other tools allowing the scientist to construct OTUs, such as MOTHUR [SCH 09], and there is a diversity of distance measurements to perform sequence clustering in OTUs [NGU 16]. For the alpha diversity, i.e. the diversity quantification within each sample, there are not only numerous indexes that can be used, including the Shannon, Simpson and Chao1 indexes, but also other measurements based on phylogeny, such as Balance-Weighted Phylogenetic Diversity (BWPD), in order to characterize diversity within samples. In the same way, there are different distance measurements for the beta diversity (Jaccard, Euclidean and Bray–Curtis to only mention three) and different representations (PCoA, NMDS, etc.). Therefore, there is indeed a path dependency for microbiota analyses; however, if stages and analyses are systematically the same, there are several ways to perform these stages. This path dependency is framed by not only standard answers but also the questions being asked, which are determined and identical in most studies (who is there? In what proportions? What is the taxonomic diversity?).

What can be learned from this categorization? If phase II makes some generalizations possible, it does not maximize standard comparisons.

Consequently, to see further, namely to extend comparative studies, one path could be to ensure that microbiome study goes from phase II to phase III. For its part, phase III imposes some analytical blinkers. Moreover, to see further, i.e. to see something else and differently, an additional path could also be applied, which would aim at making microbiota studies go from phase III to phase IV. Since, at an epistemic level, phase IV is probably not very different from phase II, or even phase I (as the difference between these phases is simply the chronological appearance order in the history of sciences) [KRO 12], these two inspirations to reinforce the holistic abilities of metagenomics first seem philosophically contradictory. We will quickly discuss the advantages and limits of each of these strategies, in order to check whether they could actually play the part expected: increasing the depth and scope of metagenomic explanations.

5.4. Standardizing metagenomics

Standardizing the production and analysis of metagenomic data sets seems in principle to provide a means to see beyond a specific metagenomic data set. In fact, different studies would then become comparable. This type of approach is especially undertaken by pooling bioinformatic scripts. This is illustrated by the case of the script enabling the location of enterotypes, which is available with a tutorial (see enterotype.embl.de). This tutorial helps to obtain a standardized method to search enterotypes and therefore make studies more comparable to each other (while remembering that the sequencing method, the type of reads, etc. can have an impact on the result obtained). Scientists were able to use this tutorial to compare their results [MOE 15, MOE 12, LIM 14, LI 17] to those obtained in the original article [ARU 11]. This makes it possible to compare achieved groups with the same clustering method (here, the PAM method of the R "cluster" package). The result of such an analysis is remarkable but ambivalent. On the one hand, we can test the existence of similar communities in different host lineages and thus give a phylogenetic interpretation of microbiome similarities associated with related species. For example, primates are supposed to share three enterotypes. Because these enterotypes are found in chimpanzees, gorillas and humans, it seems legitimate to suggest that these three types of communities could have evolved from their last common ancestor (and maybe even before). Nevertheless, and this well illustrates the difficulty or even the impossibility of actually standardizing metagenomic studies, two individuals can have the same enterotype and nonetheless host microbes with very different gene contents; simply, the same enterotypes can correspond to very different microbes. There is here a remarkable paradox explained, on the one hand, by the biological processes affecting microbial genomes and, on the other hand, by the reduction (which is slightly essentialistic) imposed by the standardization of microbial community studies. In fact, it can be said that two individuals share the same enterotype if they have the same combinations (and according to the same relative abundances) of 16S and 18S OTUs. However, microbe genomes with 16S belonging to the same OTU can already be different in significant proportions, as strains rapidly gain and lose genes [DOO 10]. Therefore, there is to a large extent a semblance (or even illusion) of comparability between standardized metagenomic data sets: sequences revealing the presence of the same OTUs in two data sets can mean that one is dealing with the "same species, except their genes!" in these two data sets (which makes it in fact impossible to rely only on 16S/18S studies in order to really understand the diversity and functioning of these communities).

On the other hand, if the bioinformatic production and analysis of data can be standardized, the very nature of the subjects studied in metagenomics and the scientific objective of this approach, namely understanding the causes of environmental diversity, mean that scientists are going to take an interest in different environments, and this will, by definition, prevent them from standardizing the various contexts from which the samples come. It is one of the main differences between laboratory studies, in which the different genetic and environmental parameters can be controlled to generate repeatable experiments, and metagenomic studies. The microbiomes available may be obtained and analyzed in a similar way, they will nonetheless come from different conditions, and comparing them will amount to comparing samples obtained, all other things not being equal. In addition, it will be very difficult to separate the impact of the different environmental parameters on the diversity of microbial communities, because it is generally complicated to maintain some of these parameters constant while varying others. Consequently, it seems realistic to foresee the discovery of correlations between different environmental factors and microbial diversity, rather than the discovery of the causes of this diversity, even after performing extremely standardized comparisons between microbiomes. Thus, several studies aim at comparing the impact of a Western human diet against a different diet (for example, a diet from Venezuela, Malawi [YAT 12], Greenland, Burkina Faso [DEF 10], etc.). The problem is that it is difficult to know whether the difference observed between these microbiomes is due to a difference in diet or the host genetic constraint (because one works on different populations), or even due to the environment (for example, present pathogens are not the same), or a combination of these variables. We could imagine seeing more clearly by also sequencing the metagenome of these organisms' environment, if it helped to check which bacteria and genes are present (and are possibly found in their intestinal microbiomes). However, such a further study would be costly and require a large storage space. Similarly, the comparative study of population genetics, whose microbiome is studied, could improve interpretations. Therefore, it is not certain that standardizing metagenomics research meets *in fine* the objectives expected by the scientists involved in this endeavor to create new path dependencies. In addition, this path would probably have the disadvantage of channeling research on diversity.

5.5. Unlocking metagenomics

Another strategy to try to widen our view in metagenomics could be to adopt the opposite attitude and fully implement a science driven by its data [LEO 12]. It would then be about multiplying data sets as well as *explananda* (the descriptions of these data and the phenomena that can be detected), in particular by unlocking "data-mining" methods to study microbiota (and microbiomes). A trivial manner to start unlocking microbiota studies consists in not only relying on 16S and 18S molecules (or on the same regions of these markers), but also considering other regions and other markers, such as ribosomal proteins, obtained during the microbiome random sequencing in order to compare the predictions (which are often contradictory) [STO 10] made by these different types of markers in respect of the taxa present in the community. An even more effective unlocking is possible. It would involve promoting methods with more exploratory than immediate pragmatic goals, i.e. methods to produce and analyze data that are selected less because of the established questions they allow one to answer than because of the new questions they could allow to be asked.

Purely exploratory "data mining" can in particular take the form of the creation of "graph-mining" sequences using similarity networks approaches. It is indeed possible to analyze the diversity of multi-faceted microbial communities, by analyzing those of their reads, contigs, genes, genomes, taxa, the co-occurrences and interactions between these taxa, etc. For example, it is possible to build similarity networks between reads with a graph structure giving information on the microbial community biology (see Figure 5.3).

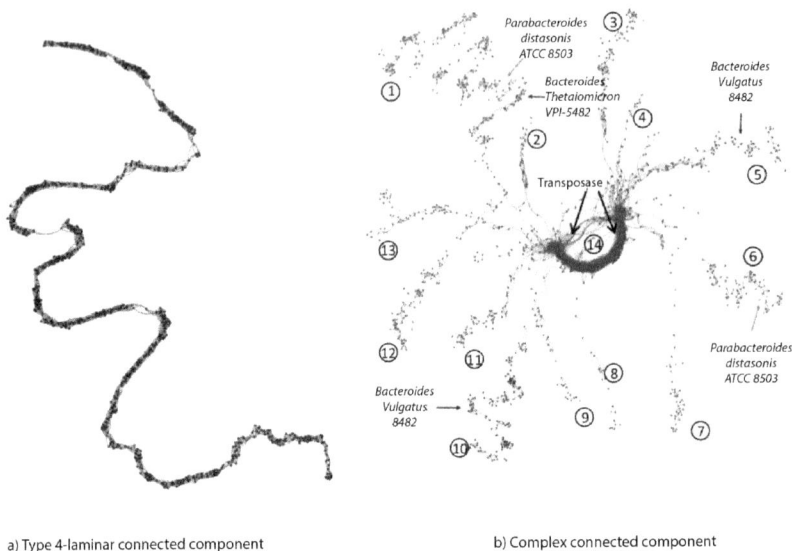

a) Type 4-laminar connected component   b) Complex connected component

Figure 5.3. *Connected components from a similarity network between reads of a lizard intestinal microbiome. For a color version of this figure, see www.iste.co.uk/grandcolas/biodiversity.zip*

In connection with the study of the impact of a diet change on the intestinal microbiome of lizards, we constructed such a similarity network between reads for each lizard. To do so, all the reads comprising each microbiome were first compared to each other, in pairs, using an all-against-all BLAST [NCB 17, ALT 90]. The output of this BLAST was filtered in order to only keep the alignments with at least 90% identity (i.e. 90% of the alignment base pairs between two reads are identical), at least 80% cover (i.e. at least one of the two reads involved in the alignment must be 80% of the length of its sequence involved in the alignment) and an E-value inferior to 1e-5. The output file is then a network registered under the form of an edge list representing similarities between two reads (which are the network nodes). This network can then be visualized thanks to tools such as Gephi [BAS 09] or Cytoscape [CHR 05]. In the ideal case where one would have managed to capture the whole diversity of a metagenome, it would be expected that this type of network would reproduce the circular genomes of bacteria. Because we know that metagenomes are environmental samples and that we cannot sequence its whole diversity, we rather expect to obtain contigs in the actual data sets, i.e. more or less long read chains, each comprising a sub-network, which is also called a connected component. This is what we mostly obtain in our networks, an observation that allowed us to characterize a new class of graph called k-laminar graphs [VOL 16].

A k-laminar graph (see Figure 5.3(a)) is a connected component with each of its nodes located at a distance inferior or equal to k edges of the diametral path, i.e. the longest of the shortest paths between two nodes in the graph. This distance k to the diametral path is interesting, because the larger it is, the more there are variant reads pertaining to the same DNA region and, therefore, the more there is a genetic diversity. The value of k is a new way of quantifying genetic diversity in a metagenome.

However, some connected components of our read networks adopt more complex topologies than a k-laminar graph, as shown in Figure 5.3(b), which represents a set resembling 13 laminars (numbered from 1 to 13 in the figure) united by a central loop (Figure 5.3, #14), which seems to create a junction between these laminars. The loop at the network center contains reads whose functional annotation corresponds to a transposase. The different parts of the connected components linked to this transposase are composed of different bacterial strains according to the taxonomic annotation. This graph provides information on genetic diversity in the microbial community of a lizard and shows how some well-conserved genes at the sequence level, here the transposase, are found in different genomic contexts. At this stage, it is however completely random to hope for a positive feedback at the end of the development of such methods. Especially because they are exploratory, their usefulness is not predictable, nor is the likelihood or possibility publishing them in journals with high impact factors. By definition, science risks getting lost when breaking new grounds... but it also has the possibility of establishing unexpected knowledge. In the particular case of read networks, this different study approach of microbial community genetic diversity identified a new network class [VOL 16]. This discovery seems for now more significant for graph theoreticians than biologists. However, who can predict if in the future this additional and new link between these disciplines will not be successful in another way?

Fundamentally, the justification for metagenomics unlocking is that a better knowledge of microbial community diversity is possible, if we consider each data set as a piece of a larger jigsaw in a voluntarily integrative approach, i.e. a strategy combining a range of approaches in order to explore biological diversity [OMA 12]. In this context, each data set and each analysis represents a piece of a jigsaw whose overall picture will be revealed later. The integrative perspective seems particularly natural in metagenomics, because the biologists studying the microbiomes of animals and plants are obviously searching for more systemic models describing the dynamics of microbial communities in host communities, or even in ecosystems. Unlocking metagenomic studies is also justified by a perspectivist [CAL 12] (less ambitious) approach, but which aims at multiplying theoretical points of view, as each data set captures aspects of microbial diversity, which is, to say the least, a complex and multi-level phenomenon. Metagenomic data can perfectly play this role. Thus, the discovery in the deepest sea areas of the sequences of a new group, the Asgard archaea, which were initially represented by Loki metagenomes, has nonetheless consolidated the theories on the symbiogenetic origin of eukaryotes [SAW 15, SPA 15]. Here, the information of a metagenome was used to shed light on an independent theory of metagenomics. Furthermore, new methods are essential, because the approaches in place struggle to annotate a large number of sequences (which well illustrates that they are insufficient on their own) [SUN 13].

If the unlocking path seems to provide a wider view, it is nonetheless an optimistic bet, because nothing guarantees that the jigsaw pieces will fit together in a general picture, rather than enriching a catalog of local and naturalistic descriptions of the microbial world. In addition, unlocking will inevitably amplify the quantity of *explananda* for which standardized comparisons are impossible, which risks inciting despair in researchers hoping to stay afloat in the metagenomic knowledge flow rather than drowning.

5.6. Conclusion

Undertaking metagenomic studies is not only interesting, but also creates high levels of frustrations, as the difficulties related to promoting local discoveries or trying to draw more general conclusions are numerous. Data are very abundant, because they can be obtained in a relatively automatic manner. This type of "convenience science" ensuring a set of conformist analyses has nevertheless a limited scope. Analyzing microbiomes in the hope of extracting the deepest and most general knowledge possible is indeed a major challenge nowadays. Even if there are centralized databases like NCBI [COO 16], HMP (see http://hmpdacc.org/), EBI [MIT 16], TARA [PES 15, SUN 15, VAR 15] and MGRAST [MEY 08], as it stands, it is complicated to compare the data and conclusions of existing studies to each other. All microbiomes cannot be compared to each other because of the diversity of their hosts, their original environments, the size and nature of these data sets and the interpretation limits of related standardized ontologies. Testing hypotheses regarding microbial diversity and identifying their causes are difficult at a local level and very difficult at a more general level. Metagenomics is then still mainly a naturalistic science producing catalogs. We foresee that this status is very likely to last given the actual size and complexity of microbiomes. We think that the (epistemic) overview we provided prompts the simultaneous conduct of two contradictory strategies: the standardization and unlocking of metagenomics, hoping that they will prove to be complementary. In fact, this pluralistic proposal is confronted with a last unavoidable practical problem in diversity studies: Which resources should and could be respectively allocated to these two strategies and to the data storage resulting from it, in order to facilitate their harmonious coexistence?

5.7. Acknowledgements

Eric Bapteste was financially supported by the ERC (European Research Council FP7/2007-2013 Grant Agreement 615274) and Chloé Vigliotti by LabexBCDIV.

The authors thank the team that collected the lizards in Croatia, especially Beck Wehrle and Anthony Herrel, as well as the company that sequenced their microbiota and microbiomes: MrDNA (Scot Dowd).

Finally, the authors thank the graph theoreticians with whom they worked to study similarity networks in lizard microbiomes: Michel Habib, Léo Planche and Finn Völkel.

5.8. Figure legends

5.8.1. *Figure 5.1. Impact of the sequencing method on the taxonomic composition of lizard microbiota*

This figure is a lizard's (X) bacterial phyla matrix representing the abundance of each phylum within each lizard microbiome, which is estimated by quantifying the number of V4-region reads of the 16S RNA assigned to this phylum. Each matrix pair (lizard X, phylum Y) was associated with a color taking into account the abundance of the phylum Y in the microbiota of lizard X. The darker the color, the more the phylum Y is abundant in the microbiota of lizard X. The three phyla in boxes correspond to the three most abundant phyla (majority phyla).

A clustering based on the Bray–Curtis distance for grouping lizards with close abundances in phyla was applied to this matrix. When we focus on the genetic and ecological characteristics of lizards within the groups, we realize that they are grouped by sequencing year. The diet of lizards was also indicated in this matrix, and a distinction between insular and continental lizards was taken into account. The sequencing method (technological constraint) has a more significant impact on the microbiota structure than the difference in diet (which is a biological reality) or a potential insularity effect.

5.8.2. *Figure 5.2. Hypothetical sequence of the stages making it possible to acquire knowledge in a scientific discipline, adapted from Sydow et al. [SYD 09]*

The x-axis represents the time axis, from the beginning of a method (on the left) to its subsequent developments (toward the right). The triangles represent the different approaches/methods used in the discipline. The lighter the triangle color, the less the method/approach it represents is used. The spread of the triangles represents the difference between methods: very similar methods occupy a near space. The key stages of the discipline evolution are indicated at the top and bottom of the figure.

5.8.3. *Figure 5.3. Connected components from a similarity network between reads of a lizard intestinal microbiome*

The network is visualized using the Gephi tool. The nodes of each network-connected component are individual reads, and two reads are linked by an edge if they have a mutual cover percentage higher than 80%, an identity percentage higher than 90% and a BLAST E-value less than 1^e-5.

a) Type 4-laminar connected component (i.e. the farthest node from the diametral path is at four edges of this path) containing 2,136 nodes and 13,991 edges. *b) More complex connected component* containing 1,843 nodes and 25,022 edges. The numbers 1–14 correspond to the different parts of the connected component. The first 13 numbers are connected component regions of taxonomically annotated laminars linked to each other by a region (loop, number 14) at the network center containing reads with a functional annotation corresponding to a transposase. The structure of this connected component (of laminars connected by a loop) provides information on microbial community biology and helps to visualize how some genes (here, the transposase) can be present in different genomic contexts (here, in populations of *Bacteroides vulgatus*, *Bacteroides thetaiotaomicron* and *Parabacteroides distasonis*).

5.9. Bibliography

[ALT 90] ALTSCHUL S., GISH W., MILLER W. *et al.*, "Basic local alignment search tool", *Journal of Molecular Biology*, no. 215, pp. 403–410, 1990.

[ARU 11] ARUMUGAM M., RAES J., PELLETIER E., *et al.*, "Enterotypes of the human gut microbiome", *Nature*, no. 473, pp. 174–180, 2011.

[BAS 09] BASTIAN M., HEYMANN S., JACOMY M., "Gephi: An open source software for exploring and manipulating networks", *International AAAI Conference on Weblogs and Social Media*, pp. 361–362, San Jose, USA, 2009.

[BOI 12] BOISVERT S., RAYMOND F., GODZARIDIS É. *et al.*, "Ray Meta: scalable de novo metagenome assembly and profiling", *Genome Biology*, no. 13, p. R122, 2012.

[BRO 15] BROWN C.T., HUG L.A., THOMAS B.C. *et al.*, "Unusual biology across a group comprising more than 15% of domain bacteria", *Nature*, no. 523, pp. 208–211, 2015.

[CAL 12] CALLEBAUT W., "Scientific perspectivism: A philosopher of science's response to the challenge of big data biology", *Studies in History and Philosophy of Science Part C*, no. 43, pp. 69–80, 2012.

[CAP 10] CAPORASO J.G., KUCZYNSKI J., STOMBAUGH J. *et al.*, "QIIME allows analysis of high-throughput community sequencing data", *Natural Methods*, no. 7, pp. 335–336, 2010.

[CHR 05] CHRISTMAS R., AVILA-CAMPILLO I., BOLOURI H. *et al.*, "Cytoscape: a software environment for integrated models of biomolecular interaction networks", *Genome Research*, no. 13, no. 11, pp. 2498–2504, 2005.

[COO 16] COORDINATORS N.R., "Database resources of the National Center for Biotechnology Information", *Nucleic Acids Research*, no. 44, pp. D7–D19, 2016.

[DEF 10] DE FILIPPO C., CAVALIERI D., DI PAOLA M. *et al.*, "Impact of diet in shaping gut microbiota revealed by a comparative study in children from Europe and rural Africa", *Proceedings of the National Academy of Sciences of the United States of America*, no. 107, pp. 14691–14696, 2010.

[DOO 10] DOOLITTLE W.F., ZHAXYBAYEVA O., "Metagenomics and the units of biological organization", *Bioscience*, no. 60, pp. 102–112, 2010.

[GIL 06] GILL S.R., POP M., DEBOY R.T. *et al.*, "Metagenomic analysis of the human distal gut microbiome", *Science*, no. 312, pp. 1355–1359, 2006.

[GOM 15] GOMEZ A., PETRZELKOVA K., YEOMAN C.J. *et al.*, "Gut microbiome composition and metabolomic profiles of wild western lowland gorillas (Gorilla gorilla gorilla) reflect host ecology", *Mol Ecol*, no. 24, pp. 2551–2565, 2015.

[HER 08] HERREL A., HUYGHE K., VANHOOYDONCK B. *et al.*, "Rapid large-scale evolutionary divergence in morphology and performance associated with exploitation of a different dietary resource", *Proceedings of the National Academy of Sciences of the United States of America*, no. 105, pp. 4792–4795, 2008.

[HON 11] HONG P-Y., WHEELER E., CANN IKO., *et al.*, "Phylogenetic analysis of the fecal microbial community in herbivorous land and marine iguanas of the Galápagos Islands using 16S rRNA-based pyrosequencing", *ISME J*, no. 5, pp. 1461–1470, 2011.

[HUM 12] HUMAN MICROBIOME PROJECT CONSORTIUM, "Structure, function and diversity of the healthy human microbiome", *Nature*, no. 486, pp. 207–214, 2012.

[JOU 17] JOUSSET A., BIENHOLD C., CHATZINOTAS A. *et al.*, "Where less may be more: how the rare biosphere pulls ecosystems strings", *ISME J*, no. 11, pp. 853–862, 2017.

[KOH 13] KOHL KD., CARY TL., KARASOV WH. *et al.*, "Restructuring of the amphibian gut microbiota through metamorphosis", *Environ Microbiol Rep*, no. 5, pp. 899–903, 2013.

[KON 05] KONSTANTINIDIS K.T., TIEDJE J.M., "Genomic insights that advance the species definition for prokaryotes", *Proceedings of the National Academy of Sciences of the United States of America*, no. 102, pp. 2567–2572, 2005.

[KRO 12] KROHS U., "Convenience experimentation", *Studies in History and Philosophy of Science Part C*, vol. 43, pp. 52–57, 2012.

[KUC 12] KUCZYNSKI J., STOMBAUGH J., WALTERS WA. *et al.*, "Using QIIME to analyze 16s rRNA gene sequences from microbial communities", *Curr Protoc Microbiol.*, no. 36, pp. 10.7:10.7.1–10.7.20, 2012.

[LEC 13] LE CHATELIER E., NIELSEN T., QIN J. *et al.*, "Richness of human gut microbiome correlates with metabolic markers", *Nature*, no. 500, pp. 541–546, 2013.

[LEO 12] LEONELLI S., "Introduction: making sense of data-driven research in the biological and biomedical sciences", *Studies in History and Philosophy of Science Part C*, no. 43, pp. 1–3, 2012.

[LI 17] LI J., POWELL J.E., GUO J. *et al.*, "Two gut community enterotypes recur in diverse bumblebee species", *Current Biology*, no. 25, pp. R652–R653, 2017.

[LIM 14] LIM M.Y., RHO M., SONG Y-M. *et al.*, "Stability of gut enterotypes in Korean monozygotic twins and their association with biomarkers and diet", *Scientific Reports*, no. 4, p. 7348, 2014.

[LIU 11] LIU B., GIBBONS T., GHODSI M. *et al.*, "Accurate and fast estimation of taxonomic profiles from metagenomic shotgun sequences", *BMC Genomics*, no. 12, p. S4, 2011.

[LIU 12] LIU L., LI Y., LI S. *et al.*, "Comparison of next-generation sequencing systems", *Journal Biomed Biotechnol*, vol. 2012, no. 251364, p. 11, 2012.

[LOP 15] LOPEZ P., HALARY S., BAPTESTE E., "Highly divergent ancient gene families in metagenomic samples are compatible with additional divisions of life", *Biology Direct*, no. 10, p. 64, 2015.

[MA 12] MA B., FORNEY L.J., RAVEL J., "Vaginal microbiome: rethinking health and disease", *Annual Review of Microbiology*, no. 66, pp. 371–89, 2012.

[MAR 11] MARTINSON V.G., DANFORTH B.N., MINCKLEY R.L. *et al.*, "A simple and distinctive microbiota associated with honey bees and bumble bees", *Molecular Ecology*, no. 20, pp. 619–628, 2011.

[MCC 14] McCann J.C., Wickersham T.A., Loor J.J., "High-throughput methods redefine the rumen microbiome and its relationship with nutrition and metabolism", *Bioinform Biol Insights*, no. 8, pp. 109–125, 2014.

[MED 17] Medina-Colorado A.A., Vincent K.L., Miller A.L. *et al.*, "Vaginal ecosystem modeling of growth patterns of anaerobic bacteria in microaerophilic conditions", *Anaerobe*, no. 45, pp. 10–18, 2017.

[MEY 08] Meyer F., Paarmann D., D'Souza M. *et al.*, "The metagenomics RAST server – a public resource for the automatic phylogenetic and functional analysis of metagenomes", *BMC Bioinformatics*, no. 9, p. 386, 2008.

[MIT 16] Mitchell A., Bucchini F., Cochrane G. *et al.*, "EBI metagenomics in 2016 - an expanding and evolving resource for the analysis and archiving of metagenomic data", *Nucleic Acids Res*, no. 44, p. D595, 2016.

[MOE 12] Moeller A.H., Degnan P.H., Pusey A.E. *et al.*, "Chimpanzees and humans harbor compositionally similar gut enterotypes", *Nature Communications*, no. 3, p. 1179, 2012.

[MOE 15] Moeller A.H., Peeters M., Ayouba A. *et al.*, "Stability of the gorilla microbiome despite simian immunodeficiency virus infection", *Molecular Ecology*, no. 24, pp. 690–697, 2015.

[NAM 12] Namiki T., Hachiya T., Tanaka H. *et al.*, "MetaVelvet: an extension of Velvet assembler to de novo metagenome assembly from short sequence reads", *Nucleic Acids Res*, no. 40, pp. e155–e155, 2012.

[NCB 17] Ncbi, BLAST web site, http://blast.ncbi.nlm.nih.gov/Blast.cgi, 2017.

[NEM 11] Nemergut DR., Costello EK., Hamady M. *et al.*, "Global patterns in the biogeography of bacterial taxa", *Environ Microbiol*, no. 13, pp. 135–144, 2011.

[NGU 16] Nguyen N-P., Warnow T., Pop M. *et al.*, "A perspective on 16S rRNA operational taxonomic unit clustering using sequence similarity", *Npj Biofilms Microbiomes*, no. 2, p. 16004, 2016.

[OMA 12] O'Malley MA., Soyer OS., "The roles of integration in molecular systems biology", *Studies in History and Philosophy of Science Part C*, no. 43, pp. 58–68, 2012.

[PEN 12] Peng Y., Leung HCM., Yiu SM. *et al.*, "IDBA-UD: a de novo assembler for single-cell and metagenomic sequencing data with highly uneven depth", *Bioinformatics*, no. 28, p. 1420, 2012.

[PES 15] PESANT S., NOT F., PICHERAL M. *et al.*, "Open science resources for the discovery and analysis of Tara Oceans data", *Science data*, vol. 2, p. 150023, 2015.

[PRA 17] PRADO-IRWIN S.R., BIRD A.K., ZINK AG. *et al.*, "Intraspecific variation in the skin-associated microbiome of a terrestrial salamander", *Microbial ecology*, pp. 1–12, 2017.

[SAW 15] SAW J.H., SPANG A., ZAREMBA-NIEDZWIEDZKA K. *et al.*, "Exploring microbial dark matter to resolve the deep archaeal ancestry of eukaryotes", *Nature*, no. 499, pp. 431–437, 2015.

[SCH 09] SCHLOSS P.D., WESTCOTT S.L., RYABIN T. *et al.*, "Introducing mothur: open-source, platform-independent, community-supported software for describing and comparing microbial communities", *Apply Environ Microbiol*, no. 75, pp. 7537–7541, 2009.

[SCH 17] SCHUELLER K., RIVA A., PFEIFFER S. *et al.*, "Members of the oral microbiota are associated with IL-8 release by gingival epithelial cells in healthy individuals", *Front Microbiol*, no. 8, p. 416, 2017.

[SEG 12] SEGATA N., WALDRON L., BALLARINI A. *et al.*, "Metagenomic microbial community profiling using unique clade-specific marker genes", *Nature Methods*, no. 9, pp. 811–814, 2012.

[SHA 14] SHARPTON T.J., "An introduction to the analysis of shotgun metagenomic data", *Front Plant Science*, no. 5, p. 209, 2014.

[SNE 62] SNEATH P.H., SOKAL R.R., "Numerical taxonomy", *Nature*, no. 193, pp. 855–860, 1962.

[SPA 15] SPANG A., SAW J.H., JORGENSEN S.L. *et al.*, "Complex archaea that bridge the gap between prokaryotes and eukaryotes", *Nature*, no. 521, p. 173–179, 2015.

[STO 10] STOECK T., BASS D., NEBEL M. *et al.*, "Multiple marker parallel tag environmental DNA sequencing reveals a highly complex eukaryotic community in marine anoxic water", *Molecular Ecology*, no. 19, pp. 21–31, 2010.

[SU 16] SU L., YANG L., HUANG S. *et al.*, "Comparative gut microbiomes of four species representing the higher and the lower termites", *Journal of Insect Science*, no. 16, p. 97, 2016.

[SUN 13] SUNAGAWA S., MENDE D.R., ZELLER G. *et al.*, "Metagenomic species profiling using universal phylogenetic marker genes", *Natural Methods*, no. 10, pp. 1196–1199, 2013.

[SUN 15] SUNAGAWA S., COELHO L.P., CHAFFRON S. *et al.*, "Structure and function of the global ocean microbiome", *Science*, vol. 348, p. 1261359, 2015.

[SYD 05] SYDOW J., SCHREYÖGG G., KOCH J., "Organizational paths: Path dependency and beyond", *21st EGOS Colloquium*, June 30–July 2, Berlin, 2005.

[TRE 11] TREANGEN T.J., KOREN S., ASTROVSKAYA I. *et al.*, "MetAMOS: a metagenomic assembly and analysis pipeline for AMOS", *Genome Biology*, no. 12, p. 25, 2011.

[TRE 13] TREANGEN T.J., KOREN S., SOMMER D.D. *et al.*, "MetAMOS. a modular and open source metagenomic assembly and analysis pipeline", *Genome Biology*, no. 14, pp. R2–R2, 2013.

[TUR 09a] TURNBAUGH P.J., RIDAURA V.K., FAITH J.J. *et al.*, "The effect of diet on the human gut microbiome: a metagenomic analysis in humanized gnotobiotic mice", *Science Translational Medicine*, no. 1, p. 6ra14, 2009.

[TUR 09b] TURNBAUGH P.J., HAMADY M., YATSUNENKO T. *et al.*, "A core gut microbiome in obese and lean twins", *Nature*, no. 457, pp. 480–484, 2009.

[VAR 15] DE VARGAS C., AUDIC S., HENRY N. *et al.*, "Eukaryotic plankton diversity in the sunlit ocean", *Science*, no. 348, pp. 1261605–1261605, 2015.

[VOL 16] VÖLKEL F., BAPTESTE E., HABIB M. *et al.*, "Read networks and k-laminar graphs", *arXiv*, 1603.01179, 2016.

[WAL 11] WALTER J., LEY R., "The human gut microbiome: ecology and recent evolutionary changes", *Annual Reviews of Microbiology*, no. 65, pp. 411–429, 2011.

[WAN 17] WANG Y., HATT J.K., TSEMENTZI D. *et al.*, "Quantifying the importance of the rare biosphere for microbial community response to organic pollutants in a freshwater ecosystem", *Appl Environ Microbiol.*, no. 8, pp. e03321–16, 2017.

[WEI 15] WEI F., WANG X., WU Q., "The giant panda gut microbiome", *Trends Microbiol.*, no. 23, pp. 450–452, 2015.

[WU 08] WU M., EISEN J.A., "A simple, fast, and accurate method of phylogenomic inference", *Genome Biol*, no. 9, pp. R151–R151, 2008.

[WU 11] WU G.D., CHEN J., HOFFMANN C., BITTINGER K. *et al.*, "Linking long-term dietary patterns with gut microbial enterotypes", *Science*, no. 334, pp. 105–108, 2011.

[WU 12] WU M., SCOTT AJ., "Phylogenomic analysis of bacterial and archaeal sequences with AMPHORA2", *Bioinformatics*, no. 28, p. 1033, 2012.

[XU 16] Xu J., GALLEY J.D., BAILEY M.T. *et al.*, "The impact of dietary energy intake early in life on the colonic microbiota of adult mice", *Scientific Reports*, no. 6, p. 19083, 2016.

[YAN 15] YÁÑEZ-RUIZ D.R., ABECIA L., NEWBOLD C.J., "Manipulating rumen microbiome and fermentation through interventions during early life: A review", *Front Microbiol.*, no. 6, p. 1133, 2015.

[YAT 12] YATSUNENKO T., REY F.E., MANARY M.J. *et al.*, "Human gut microbiome viewed across age and geography", *Nature*, no. 486, pp. 222–227, 2012.

[ZEN 15] ZENG B., HAN S., WANG P. *et al.*, "The bacterial communities associated with fecal types and body weight of rex rabbits", *Scientific Reports*, no. 5, p. 9342, 2015.

[ZHE 16] ZHENG J., XIAO X., ZHANG Q. *et al.*, "The programming effects of nutrition-induced catch-up growth on gut microbiota and metabolic diseases in adult mice", *Microbiologyopen*, no. 5, pp. 296–306, 2016.

[ZHU 11] ZHU L., WU Q., DAI J. *et al.*, "Evidence of cellulose metabolism by the giant panda gut microbiome", *Proceedings of the National Academy of Sciences of the United States of America*, no. 108, pp. 17714–17719, 2011.

Genetic Code Degeneracy and Amino Acid Frequency in Proteomes

6.1. Introduction

Proteins assembled by the cell are made up of approximately 20 different kinds of amino acids which are encoded in the genome in accordance with the rules of the genetic code. The structure of each expressed protein, which can count up to several thousand amino acids, is the result of a long evolution during which the composition in amino acids has more or less changed following mutations within the coding sequences and the adaptation of mutants. Thus, the composition in amino acids of a protein reflects the requirements of its specific function. Therefore, we could speculate that no statistical trend should emerge from the analysis of proteomes, which comprise all the proteins of an organism, a genus or even all living beings.

By analyzing proteins with known amino acid sequences that were available at the end of the 1960s, King and Jukes [KIN 69] noted that the most represented amino acids were encoded by a larger number of codons in the canonical genetic code table (in other words, they were characterized by a higher degeneracy). Leucine, for example, which is on average the most frequent amino acid in proteins, is encoded by six different codons (TTG, TTA, CTT, CTC, CTG and CTA). At the other extreme, tryptophan is only encoded by a single type of codon (TGG), and its global frequency is the lowest among the 20 amino acids. This observation prompted them to

Chapter written by Jean LEHMANN.

establish the frequency of amino acids in polypeptides obtained through the translation of random sequences constructed with the same proportions of the four nucleic bases as the ones coding for these proteins. By comparing the amino acid frequencies of these two sets, they noted that they were highly correlated (Figure 6.1). This analysis shows that the *average* frequency of the amino acids in proteins is, in a first approximation, *the reflection of the genetic code degeneracy*. The discrepancies observed, several of which are significant, indicate that evolutionary forces modifying the frequency of some amino acids can be detected at the level of proteomes.

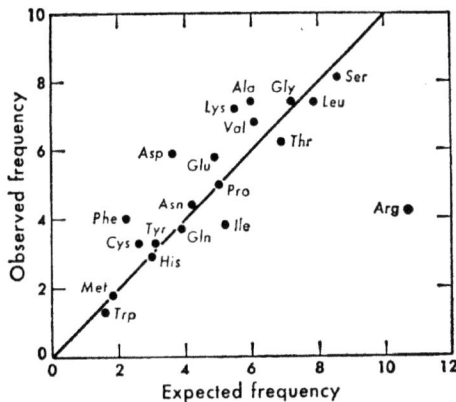

Figure 6.1. *Similarity between the observed amino acid frequencies of 53 mammal proteins and the frequencies obtained through the translation of random sequences characterized by the same nucleotidic frequencies. Reproduced from King and Jukes [KIN 69]. © American Association for the Advancement of Science 1969*

As emphasized by King and Jukes [KIN 69], this correlation shows that the living world has, on the basis of genetic drift, the capacity to generate functional proteins with amino acids whose relative proportions in these sequences are specified by the genetic code. One of their important conclusions was that many mutations are neutral from the point of view of evolution. This observation also shows that the structure of the code seems relatively optimal to generate proteins through genetic drift, without evolutionary forces having to significantly correct amino acid frequencies.

The observation of King and Jukes was based on a limited amount of data. The very large amount of sequences identified since then (the data bank *Genbank* reached 200 million recorded sequences in 2017) confirmed this fact [GRE 13]. A significant feature of the distribution was more recently identified (Figure 6.2): the frequency of the amino acids negatively correlates with their mass (or their volume) [GRE 13, LEH 16]. As the genetic code is mainly characterized by 2x and 4x degenerate codon families (see section 6.4), small amino acids, such as glycine, are thus most often encoded by 4x degenerate families, while amino acids composed of a large number of atoms are generally encoded by 2x degenerate families. The case of exceptionally encoded amino acids, such as selenocysteine, is not discussed here.

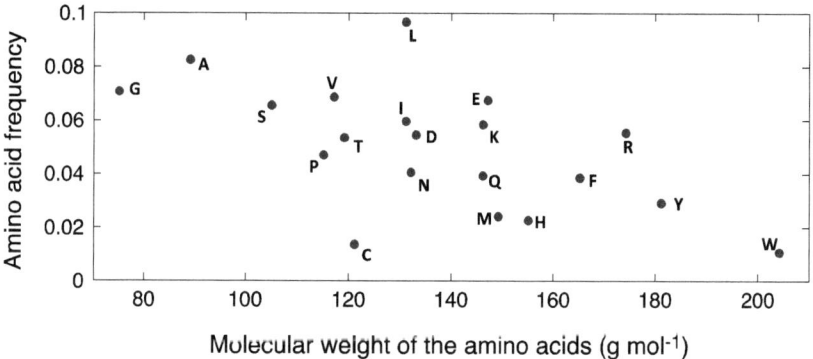

Figure 6.2. *Frequency of the encoded amino acids as a function of their mass (one letter code). The Spearman correlation coefficient is −0.6032 (p = 0.0049). Figure established from the set of sequenced proteins available in the UniProtKB/ Swiss-Prot database. Adapted from [LEH 16] © Elsevier 2016*

What is the origin of the genetic code degeneracy and that of the amino acids frequency–mass correlation? In this chapter, we will see that two different phenomena, linked by the anticodon–codon interaction, have combined during the evolution of the translation system, which resulted in the assignment of 4x degenerate codon families to small amino acids, while larger amino acids became mostly assigned to 2x degenerate families. Additional principles need to be relied on for the other levels of degeneracy, in particular for serine, leucine and arginine, as each of these amino acids are encoded by two codon families.

6.2. Frequency–mass correlation of encoded amino acids

It is important to realize that the correlation in Figure 6.2 shows a property that is not due to the evolutionary history of specific organisms, since it is established on the basis of the proteomes of an exhaustive set of organisms as different as a bacterium and a tree. It reveals a fundamental aspect of the genetic code structure and informs us about the functional needs of proteins as a whole.

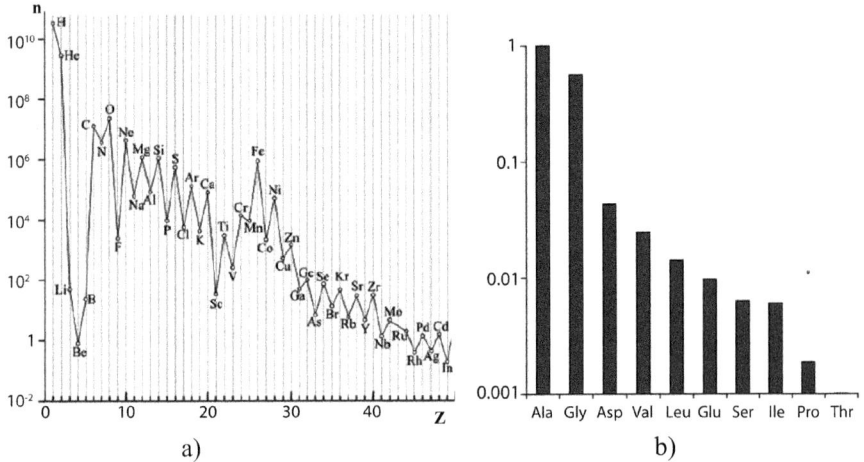

a)   b)

Figure 6.3. *a) Relative abundance of the elements in the universe as a function of the atomic number Z (up to Z = 50). b) Relative abundance of the amino acids of the genetic code in a prebiotic chemistry experiment. Figure established from Miller's Table 3 [MIL 87]*

This distribution reminds us a bit that of the elements in the universe, where some of the lightest atoms (such as hydrogen) are very abundant and appeared very early in the cosmic evolution, while heavy elements (such as copper) are far rarer, and could only be generated at the ends of the lives of the first stars [REE 81]. However, frequency scales are not comparable: one is linear (Figure 6.2) whereas the other is logarithmic (Figure 6.3(a)). The analogy with the elements is certainly more relevant if we consider the ease with which amino acids can be generated under conditions thought to reproduce those of the Earth 4 billion years ago. The famous experiments of Miller and Urey (Figure 6.3(b)) showed that the two smallest amino acids, glycine and alanine, are obtained in concentrations of at least one order of magnitude higher than the following amino acids, in the order of abundance. At the other extreme, the most complex amino acids of the genetic code, such as lysine or arginine, could not be revealed in this type of experiment [MIL 87].

Although the criterion of the relative frequency of the amino acids certainly plays a fundamental role at the origin of the genetic code [LEH 09], we will see that the physico-chemical reason for the distribution observed in the proteomes of modern organisms is not connected to amino acid abundance. It is due to the greater or lesser ease with which amino acids loaded onto transfer RNAs could form the peptide bond in the absence of a catalytic site on ancestral ribosomes, a constraint that contributed to implementing the genetic code. This hypothesis is supported by the existence of a correlation between the stability of anticodon–codon interactions and the volume of encoded amino acids (see the following section).

6.3. Amino-acid volume correlation in the genetic code

The rules of the genetic code did not fall into place randomly. When the canonical code was completely clarified around 1965, it soon appeared that the hydrophobicity of amino acids and that of nucleic bases had played an important role during its formation [WOE 66, LAC 83]. It is according to this property that Jungck [JUN 78] had suggested representing the code table. Thus, the succession [A, G, C, U], which ranks the nucleic bases following their decreasing hydrophobicity, makes it possible to group the amino acids in the table in a remarkably consistent way due to a hydrophobicity correlation between anticodons and amino acids [LAC 83]. The opposite order [U, C, G, A] also shows this consistency. For some obscure reason, the succession [U, C, A, G], which does not reveal any specific order in the table, is still often used nowadays. A second physico-chemical correlation connecting codons to amino acids was more recently identified (Figure 6.4(a)) and shows that the stability of complementary anticodon–codon pairs (ΔG_0 anticodon–codon) is stronger when the amino acid has a small side-chain [LEH 00].

As we will be demonstrating, this correlation is one of the two jigsaw pieces at the origin of the frequency–mass correlation in proteomes (Figure 6.2). In Figure 6.4(a), the amino acid size is expressed with the van der Waals volume. The related correlation coefficient is better than that obtained with mass (not shown here), suggesting a physical explanation based on steric hindrance. The energies of the complementary anticodon–codon interactions were calculated with thermodynamic parameters established from RNA strands' association in solution [TUR 87]. This point is significant, because no predefined secondary structure *a priori* characterizes these strands. Yet, anticodon loops of modern transfer RNAs all have a U-turn structure, which promotes their association with codons. The U-turn is mediated by the U33 base, which is extremely conserved [AUF 01], and this structure also often requires the presence of modified bases in the loop [GRO 98, VEN 08]. Thus, the nature of the thermodynamic parameters used to establish the correlation indicates that we are dealing with transfer RNA loops which are likely unstructured and deprived of modification. We can then reasonably conclude that this correlation reflects a physico-chemical constraint at an elementary stage of the genetic system.

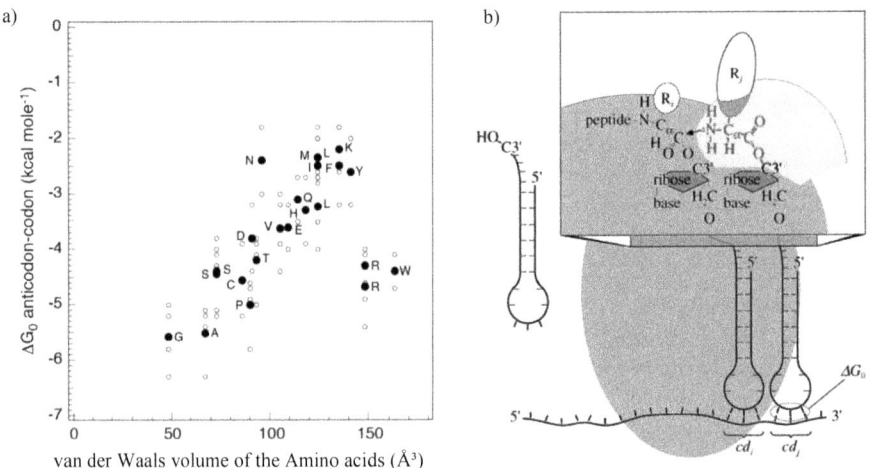

Figure 6.4. *a) Correlation in the genetic code between the van der Waals volume of amino acids and the Gibbs free energy change (ΔG_0) of anticodon–codon interactions, as estimated from the parameters of Turner et al. [TUR 87]. Small circles correspond to complementary interactions, while black dots represent averages over four codons (for 4x degenerate families) or two codons (for 2x degenerate families). b) Correlation interpretation. In an elementary translation system, the ΔG_0 of anticodon–codon interactions determines the characteristic residence time τ of a transfer RNA on its complementary codon (which can be linked to a dissociation rate constant $k-$ = $1/\tau$). In the absence of amino acid confinement on the ancestral ribosome (ellipse in gray), the van der Waals volume of side chains has a significant influence on the conformational properties of the backbone of the amino acyl (in cyan), and thus on the kinetic constant of peptide bond formation **kcat**: amino acids with a large side chain have a faster kinetics than small ones. The correlation reflects a situation where **kcat** ≈ **k**– for all (amino acid\codon) couples. Adapted from [LEH 00], © Elsevier 2000. For a color version of this figure, see www.iste.co.uk/grandcolas/biodiversity.zip*

If we connect the ΔG_0 of these interactions to the dissociation rate constant ($k-$) of transfer RNAs, which determines the average residence time of these RNAs on their complementary codons, the correlation leads us to wonder whether the kinetic constant of peptide bond formation (*kcat*) is faster for large amino acids compared to small ones [LEH 00]. Data on intramolecular reactions are fully consistent with this possibility (see, for example, [LIG 96, JUN 05]) and show that an addition of atoms (or groups of atoms) onto a carbon close to a terminal nucleophile can increase the kinetics by up to several orders of magnitude [ARM 03]. Remarkably, the correlation suggests a *monotonic* increase of the *kcat* constants with the size of the side-chains; a phenomenon that typically characterizes intramolecular systems, for which the kinetics are established in solution [LIG 96, JUN 05]. Yet, the modern ribosome confines the amino acids inside its catalytic site (the peptidyl transferase center), which significantly reduces the effect of the side chains and ensures the near standardization of the reaction rates [LEH 17]. Here again, this element suggests that the correlation reflects a physico-chemical constraint at an elementary stage, when the function of the ancestral ribosome was probably limited to the stabilization of the elongation complex (Figure 6.4(b)). It shows that an optimization phenomenon that selected the systems for which all (amino acid|codon) couples verify the equality *kcat* $\approx$ *k-* occurred with polymerization, thus contributing to the implementation of the genetic code [LEH 00, LEH 09].

An examination of the correlation shows that some (amino acid|codon) couples stand apart from the main sequence, especially asparagine (N), arginine (R) and tryptophan (W). Should other principles be called upon to explain these apparent discrepancies? If we exclusively focus on chemical reactivity, it is also necessary to consider the pk_a of the amine of the 20 amino acids. The amine group of asparaginyl-tRNA in solution is characterized by the lowest pk_a value ($\sim$6.8 to 20° C), which makes it a quite reactive nucleophile at physiological pH compared to the majority of the other amino acids (pk_a between 7 and 8). At the other extreme, the amine of prolyl-tRNA has a pk_a of approximately 8.6 [JOH 11]. The discrepancies observed in the correlation for these two amino acids (less pronounced in the case of proline) are consistent with the pk_a contribution, and suggest that the pH of the reactional environment at this stage of the evolution of the genetic code was already approximately 7.5. Considering arginine and tryptophan, a threshold effect connected with the size of the side-chain could explain why they stand apart from the main sequence.

Before concluding this section, it is appropriate to examine at least one gray area of the correlation. The latter is established from the genetic code, which is characterized by a codon degeneracy that is more or less extended. Yet, degeneracy as we know it is a property of *modern* genetic code; its canonical form was forged by the decoding center of the small ribosomal subunit and the U-turn structure of the anticodon loops of all transfer RNAs (see section 6.4). As pointed out earlier, the second condition is incompatible with the correlation, and also requires the participation of modification enzymes. At the evolutionary stage when the optimization of kinetic constraints implemented the genetic code, the productive anticodon–codon interactions could not correspond to all the Watson–Crick type pairing indicated in small circles in Figure 6.4(a), because this would involve more than 60 different tRNAs (in other words, a greater complexity than the current system) with an impossibility *per se* for the ribosome to ensure the accuracy of these pairings [EHR 80]. It is thus likely that a restricted number of codons constituted the sequences of coding RNAs, and that a limited set of transfer RNAs decoded these RNAs. In this scenario, more or less extended "wobble" types of pairings already gave rise to a certain form of degeneracy. Unlike with the modern genetic code, the contours of this degeneracy must have been blurred, with coding ambiguities between similar codons.

6.4. Origin of genetic code degeneracy

In section 6.3, we saw that the stability of the anticodon–codon interaction played a major role in the establishment of the genetic code at a stage when the ancestral ribosome did not have a catalytic site and a decoding center (Figure 6.5(a)). The latter, which tests the anticodon–codon pairing geometry, requires the U-turn structure on all transfer RNAs.

An analysis [LEH 08] established that the stability of the anticodon–codon interaction also implemented the degeneracy of the genetic code. In that case, it is more specifically the stability of the central Watson–Crick base pair that is decisive. This stability depends on the number of hydrogen bonds involved in the Watson–Crick base pairs in position 1 and 2 of the codons, to which a contribution from U_{33} may occur. This nucleotide is responsible for structuring the U-turn in the loop (Figure 6.5(b)). It can form a stable hydrogen bond with the base at the center of the anticodon (N35) if it is a purine [AUF 01]. The case of the mitochondrial system, for which the number of transfer RNAs necessary for decoding all the codons is at a minimum [BON 80], helps to explain degeneracy. This property originates from the tolerance of non-Watson–Crick base pairing at the third position of the codons. They are possible because the ribosome (almost) does not constrain the nature of this paring. However, ribosomal residues A1492, A1493 and G530 (*Escherichia coli* numbering), which form the decoding center, test the *geometrical* nature of pairing in positions 1 and 2 through specific contacts with the minor groove [OGL 01]. This test, the physical understanding of which is still debated [DEM 12, ZEN 14, ROZ 16], leads to the rejection of the transfer RNA when these two base pairs are not Watson–Crick.

In the mitochondrial system [BON 80], there are (**i**) transfer RNAs with anticodons that can base pair with any of the four codons of the 4x degenerate families and be "accepted" by the decoding center (it is referred to as "superwobbling"). The base at the 5' side of these anticodons is often an unmodified U, sometimes another base. As for the 2x degenerate families, pairing specificity must be higher: purine–purine and pyrimidine–pyrimidine pairings are excluded. Thus, (**ii**) two different transfer RNAs are required for translation of two 2x degenerate families with the same nucleotides in positions 1 and 2 (it is referred to as "simple" wobbling). The origin of the two categories (**i**) and (**ii**) lies in the stability of the base pair at the second position (N_{35}-N_2), necessarily Watson–Crick, which is conditioned by *a single* parameter: the sum of hydrogen bonds formed by the base pairs at the first and second positions, and the U_{33}-N_{35} hydrogen bond when it occurs (if N_{35} is a purine) (Figure 6.5(b)). Superwobbling is possible when there are at least six of these hydrogen bonds [LEH 08]. This phenomenon originates from the disturbance of the Watson–Crick geometry of the base pair at the second position by its neighbor at the third position (the "wobble" position). With six hydrogen bonds or more, the highest disturbances (purine–purine or pyrimidine–pyrimidine) generated by the base pair at the third position are contained because N_{35}-N_2 is sufficiently stabilized. Thus, the decoding center does not "notice" any geometrical problem at the second position, and any pairing is accepted. With five or less hydrogen bonds, the stability of N_{35}-N_2 is insufficient to contain the perturbation generated by purine–purine or pyrimidine–pyrimidine base pairs at the third position, implying that two *different* transfer RNAs are necessary to decode the four codons specified by the first two positions. It should be mentioned that base modifications are nearly always required to ensure a "clean" decoding specificity between two adjacent 2x degenerate families [LEH 08].

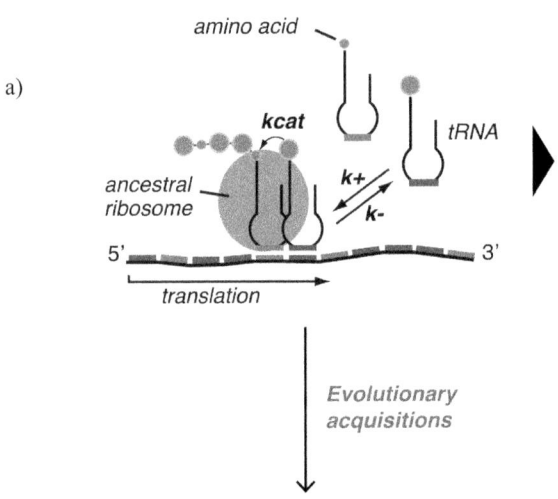

a)

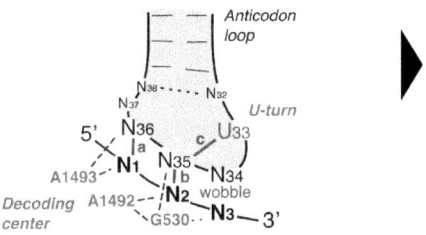

b)

Figure 6.5. *Evolution of the translation system and the genetic code. a) Elementary translation system in which the probability of peptide bond formation depends in particular on **k**– constants (determined by the anticodon–codon interaction) and the **kcat** constant (determined by the amino acid in the absence of a catalytic site on the ancestral ribosome). A partial analysis of the problem is presented in Lehmann et al. [LEH 09]. b) Anticodon–codon interaction in the decoding center of a modern ribosome. The stability of the Watson–Crick geometry of the N_{35}-N_2 base pair, which is monitored by the decoding center, is conditioned by the sum $S = a + b + c$ of hydrogen bonds. ($A_{36} \cdot U_1$ or $U_{36} \cdot A_1$: $a = 2$; $G_{36} \cdot C_1$ or $C_{36} \cdot G_1$: $a = 3$; $A_{35} \cdot U_2$ or $U_{35} \cdot A_2$: $b = 2$; $G_{35} \cdot C_2$ or $C_{35} \cdot G_2$: $b = 3$; R_{35}: $c = 1$; Y_{35}: $c = 0$). With $S \leq 5$, two transfer RNAs, with either a purine or a pyrimidine in position N_{34}, are required for translation of the four corresponding codons, which have either a purine or a pyrimidine in position N_3. With $S > 5$, a single transfer RNA can decode the four corresponding codons.*

c)

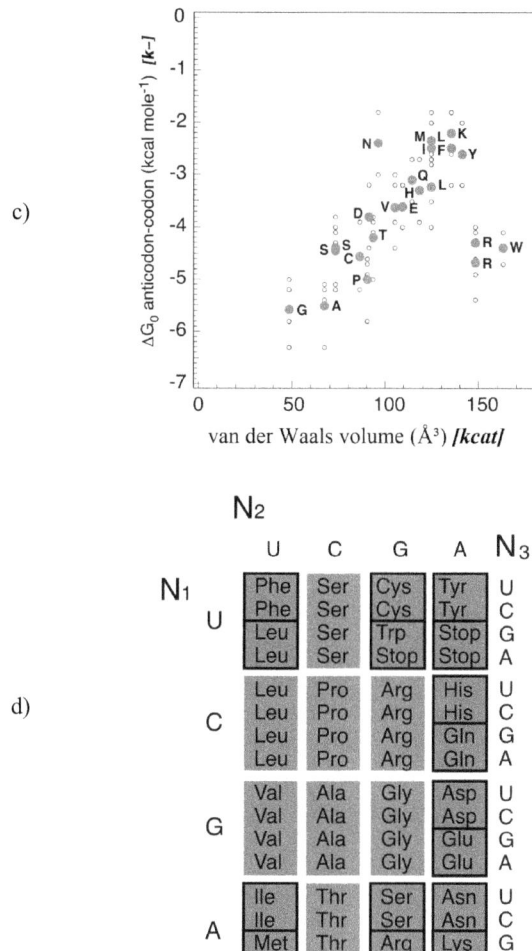

d)

c) Volume correlation (see Figure 6.4(a)), with degeneracy families indicated in blue (4x) and orange (2x). The explanatory physical quantities, **k–** and **kcat**, are indicated in brackets, although there is no continuous relation between the VDW volume and **kcat**. d) Table of the canonical genetic code with the two main degeneracy families indicated in blue (4x) and orange (2x), as determined by the analysis in (b). For a color version of this figure, see www.iste.co.uk/grandcolas/biodiversity.zip

The above analysis shows that degeneracy corresponds to the maximization of non-specific codon pairings at the third position, as it occurs in the mitochondrial system. There are two cases in the canonical code for which degeneracy is absent, i.e. where the two codons of a 2x degenerate mitochondrial family (AUR, UGR) have different assignments. Furthermore, pairing rules are usually more stringent in systems other than mitochondrion, implying that two transfer RNAs (or more) are necessary for the translation of all the codons of 4x degenerate families.

It is important to realize that the structural constraints just described are associated with a relatively *modern* genetic system (presence of a decoding center on the ribosome and U-turn in the anticodon loop of all tRNAs, which requires modifying enzymes). Moreover, the structuring of degeneracy occurred *independently* of the amino acids, as shown in Figure 6.5(b). This model led to a proposal about the set of transfer RNAs occurring at the time of the *Last Universal Common Ancestor* (LUCA) [VAN 16].

6.5. Origin of the frequency–mass correlation

Combining the elements discussed in sections 6.3 and 6.4, we can first try to extract a scenario for the evolution of the genetic code. We start from a state of the system where many actors are already present, in particular transfer RNAs, information-carrying RNAs and an ancestral ribosome (Figure 6.5(a)). The process through which initial transfer RNAs were loaded with amino acids at their 3' end is unknown, but the discovery of synthetic ribozymes capable of self-aminoacylation [ILL 95] suggests that this loading was mediated by the transfer RNAs themselves. A joint analysis of the volume (Figure 6.4) and hydrophobicity (see section 6.3) correlations is consistent with this possibility [LEH 00]. Based on a logic of structural complexification (see Figure 6.5), the following evolutionary transitions can be conjectured:

1) Although the processes through which RNA and the initial translation emerged are unknown, it has been suggested that a system constituted by sequences with the $^{5'}GNC^{3'}$ base pattern and encoding the four simplest amino acids could ensure the stability of a translation achieved by rudimentary RNA structures [EIG 78, LEH 09, WAN 17].

2) The proliferation of proto-organisms in which the kinetic constants associated with translation are optimized (Figure 6.4) sets the foundation of the genetic code. Primitive translation with the early genetic code enables

the synthesis of the first synthetases, which will extend and crystallize optimal (amino acid|anticodon) associations, and thus provide some stability to the early code.

3) With the core of the code being more or less established, the peptidyl transferase center takes shape, which leads to a standardization of the *kcat*s [LEH 17]. The optimization rule *kcat* ≈ *k*– thus vanishes.

4) The "release" of the *k*–s of the anticodon–codon pairs makes the optimization of these interactions possible through the structuring of all anticodon loops into the U-turn. This stage requires modification enzymes.

5) Appearance of the decoding center, which crystallizes degeneracy (Figure 6.5(b), (d)).

6) Appearance of mechanisms correcting decoding errors (proofreading) [IEO 16].

The origin of the frequency–mass correlation (Figure 6.2) can be understood by referring to points 2 and 5. Although the involved anticodon-codon stability is not the same in the two cases (Figure 6.5(a), (b)), it is the common factor responsible for (2) the assignment of amino acids to codons mainly according to their size and (5) the type of degeneracy of the codon families, which depends on the first two coding positions. The consequence of this overlap is shown in Figure 6.5(c): there is a clear grouping of 4x degenerate families towards small amino acids (in blue), whereas large amino acids are nearly all encoded by 2x degenerate families (in orange), as can also be seen in Figure 6.5(d). As amino acids are mostly assigned to 2x and 4x degenerate families, this phenomenon explains the frequency–mass correlation. If the frequency of the amino acids is normalized by the degeneracy characterizing them in the canonical code, the slope of the regression line becomes almost zero; the correlation thus disappears (Figure 6.6(a)).

Some amino acids are more frequently or less frequently encoded than expected from their degeneracy in the code and a random genetic drift of the sequences. An analysis of this point is presented in Greenbaum *et al.* [GRE 13], Figure S1. Cysteine, for example, is under-represented. It means that an adaptive force tends to eliminate the sequences that integrate the codons of this amino acid through mutation. It can be noted that cysteine is encoded by only two codons, i.e. a low degeneracy. As cysteine has this very special property to make disulfide bonds between specific positions (if there are any) (Figure 6.6(b)), it is not surprising that only a few residues are present in proteins, at lower doses than that determined by its degeneracy in the code. Other amino acids, such as aspartic acid, are over-represented: in that case, an adaptive force tends to keep the sequences integrating this amino acid, which is often present in catalytic sites. We can conclude by mentioning the case of leucine, an amino acid that is very present in the genetic code since it is encoded by two codon families (4x and 2x degenerate). As the normalization shows, the overall frequency of this amino acid is very close to the expected value [GRE 13]. Is the proteins' "need" for leucine the reason why two codon families ended up being associated with leucine? This point is fascinating; it suggests that this hydrophobic amino acid, much involved in protein structures such as "leucine zippers" and transmembrane regions (Figure 6.6(c)), already played a significant role when the genetic code was taking shape. Our analysis neither provides an answer to this question nor does it shed light on the phenomenon or phenomena through which serine and arginine were also assigned to two codon families in the genetic code.

6.6. Summary and discussion

The aim of the present analysis was to explain the origin of the frequency–mass correlation of the encoded amino acids as it appears when all known proteomes are being considered, all species combined. We started from the analysis of King and Jukes [KIN 69] on degeneracy, the authors of which showed that it determines in a first approximation the frequency of the amino acids in proteomes. As large amino acids have a lower degeneracy than small ones, they are globally less represented in sequences. The search for an explanation of this second phenomenon made us go back to the origin of the genetic code, where the establishment of a physico-chemical correlation (or relationship) in the code between the van der Waals volume of the amino acids and the stability of the anticodon–codon interactions laid the foundation of the frequency–mass correlation. Later in evolution, the structuring of all transfer RNA anticodon loops and the appearance of the ribosome decoding center carved the structure of degeneracy in the code (Figure 6.5(d)). This implementation, which is also based on the anticodon–codon interaction, assigned 4x degenerate families to small amino acids and 2x degenerate families to large amino acids. Some exceptions and the limitations of this analysis were discussed.

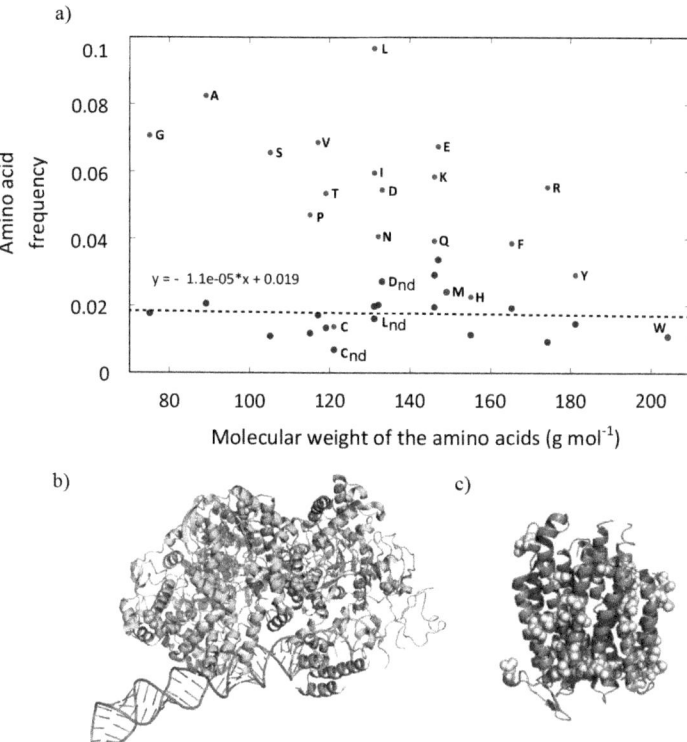

a)

$y = -1.1e-05 \cdot x + 0.019$

Figure 6.6. a) Normalization of amino acid frequencies in proteomes by their degeneracy in the canonical code. In red: before normalization (see Figure 6.2); in blue: after normalization. The slope of the regression line after normalization (indicated on the figure) is almost zero. The normalized frequencies of the amino acids discussed in the text are highlighted by C_{nd}, D_{nd} and L_{nd}. Adapted from [LEH 16], © Elsevier 2016. b) Co-crystallized RNA-dependent RNA polymerase with a double-stranded RNA (pdb 5H0R, Li et al. [LI 17]). Out of the 1,786 amino acids of this protein, 18 (=1%) are cysteines (highlighted in gold, representation with van der Waals spheres). c) Nitrate transporter NarU of E. coli (pdb 4lu9, Yan et al. [YAN 13]), illustrating a protein particularly rich in leucine: out of the 468 residues of this protein, 56 (=12.2%) are leucines (highlighted in yellow, representation with van der Waals spheres). For a color version of this figure, see www.iste.co.uk/grandcolas/biodiversity.zip

A relevant question that could be raised is why the volume and the mass of amino acids are alternatively considered throughout this chapter. The reason is circumstantial: mass is the physical quantity that was initially correlated to the amino acid frequency [GRE 13]. It turned out that the physical property explaining the size effect is more directly connected to the volume of the amino acids: in the interpretation discussed in section 6.3, we pointed out that steric effects determine the kinetic of peptide bond formation in the absence of a catalytic site on the early ribosome. To illustrate an interesting parallel with the frequency–mass distribution of the elements in the universe, mass was nevertheless kept as a variable in the correlation with the amino acid frequency. In this connection, a physical approach helps to better appreciate the similarities and differences between these two phenomena. Mass (or, if neutrons are neglected, the atomic number Z) is clearly the appropriate variable to describe the distribution of the elements, because it is connected to the energy involved in the synthesis of heavy elements from light ones. However, energy levels involved in biochemical processes are incomparably smaller. These processes rely in particular on entropic effects (the positioning of a nucleophile), for which volume is the appropriate physical quantity to describe the phenomena.

6.7. Conclusion

In this chapter, we showed that statistics on whole proteomes reveal the structure of the genetic code. At this large scale, they also reveal the driving forces of evolution: the genetic drift of coding sequences and the evolutionary forces arising from functional constraints associated with proteins. Nature managed to establish a genetic code that spontaneously provides, in accordance with functional needs, a large amount of leucine and a pinch of tryptophan. This same code provides slightly too much cysteine, and a deficit in aspartic acid, such that evolutionary forces correct so as to generate proteins necessary for the molecular dance of life.

6.8. Acknowledgments

I thank Albert Libchaber and Benjamin Greenbaum for the stimulating discussions we had on the analysis of the frequency–mass correlation, some important elements of which are included in this chapter. I also would like to thank Marie-Christine Maurel for encouragements.

6.9. Bibliography

[ARM 03] ARMSTRONG A.A., AMZEL L.M., "Role of entropy in increased rates of intramolecular reactions", *Journal of the American Chemical Society,* vol. 125, pp. 14596–14602, 2003.

[AUF 01] AUFFINGER P., WESTHOF E., "An extended structural signature for the tRNA anticodon loop", *RNA*, vol. 7, pp. 334–341, 2001.

[BON 80] BONITZ S.G., BERLANI R., CORUZZI G. *et al.*, "Codon recognition in yeast mitochondria", *Proceedings of the National Academy of Sciences of the United States of America,* vol. 77, pp. 3167–3170, 1980.

[DEM 12] DEMESHKINA N., JENNER L., WESTHOF E. *et al.*, "A new understanding of the decoding principle on the ribosome", *Nature*, vol. 484, pp. 256–259, 2012.

[EHR 80] EHRENBERG M., BLOMBERG C., "Thermodynamic constraints on kinetic proofreading in biosynthetic pathways", *Biophys Journal,* vol. 31, pp. 333–358, 1980.

[EIG 78] EIGEN M., SCHUSTER P., "The hypercycle. A principle of natural self-organization. Part C: The realistic hypercycle", *Naturwissenschaften*, no. 65, pp. 341–369, 1978.

[GRE 13] GREENBAUM B., KUMAR, P., LIBCHABER, A., "Amino acid distributions and the effect of optimal growth temperature", arXiv:1309. 4761, 2013.

[GRO 98] GROSJEAN H., HOUSSIER C., ROMBY P. *et al.*, "Modulatory role of modified nucleotides in RNA loop-loop interaction", in GROSJEAN H., BENNE R. (eds), *Modification and Editing of RNA*, ASM Press, Washington, 1998.

[IEO 16] IEONG K.W., UZUN Ü., SELMER M. *et al.*, "Two proofreading steps amplify the accuracy of genetic code translation", *Proceedings of the National Academy of Sciences of the United States of America,* vol. 113, pp. 13744–13749, 2016.

[ILL 95] ILLANGASEKARE M., SANCHEZ G., NICKLES T. *et al.*, "Aminoacyl–RNA synthesis catalyzed by an RNA", *Science*, vol. 267, pp. 643–647, 1995.

[JOH 11] JOHANSSON M., IEONG K.-W., TROBRO S. *et al.*, "pH-sensitivity of the ribosomal peptidyl transfer reaction dependent on the identity of the A-site aminoacyl-tRNA", *Proceedings of the National Academy of Sciences of the United States of America,* vol. 108, pp. 79–84, 2011.

[JUN 78] JUNGCK J.R., "The genetic code as a periodic table", *Journal of Molecular Evolution*, vol. 11, pp. 211–24, 1978.

[JUN 05] JUNG M.E., PIIZZI G. "Gem-Disubstituent effect: Theoretical basis and synthetic applications", *Chemical Reviews*, vol. 105, pp. 1735–1766, 2005.

[KIN 69] KING J.L., JUKES T.H., "Non-Darwinian evolution", *Science*, vol. 164, pp. 788–798, 1969.

[LAC 83] LACEY J.C. JR, MULLINS D.W. JR, "Experimental studies related to the origin of the genetic code and the process of protein synthesis: a review", *Origins Life*, vol. 13, pp. 3–42, 1983.

[LEH 00] LEHMANN J., "Physico-chemical constraints connected with the coding properties of the genetic system", *Journal of Theoretical Biology*, vol. 202, pp. 129–144, 2000.

[LEH 08] LEHMANN J., LIBCHABER A., "Degeneracy of the genetic code and stability of the base pair at the second position of the anticodon", *RNA*, vol. 14, pp. 1264–1269, 2008.

[LEH 09] LEHMANN J., CIBILS M., LIBCHABER A., "Emergence of a code in the polymerization of amino acids along RNA template", *PLoS One*, vol. 4, p. e5773, 2009.

[LEH 16] LEHMANN J., LIBCHABER A., GREENBAUM B., "Fundamental amino acid mass distributions and entropy costs in proteomes", *Journal of Theoretical Biology*, vol. 410, pp. 119–124, 2016.

[LEH 17] LEHMANN J., "Induced fit of the peptidyl-transferase center of the ribosome and conformational freedom of the esterified amino acids", *RNA*, vol. 23, pp. 229–239, 2017.

[LI 17] LI X., ZHOU N., CHEN W. *et al.,* "Near-Atomic Resolution Structure Determination of a Cypovirus Capsid and Polymerase Complex Using Cryo-EM at 200kV", *Journal of Molecular Biology*, vol. 429, pp. 79–87, 2017.

[LIG 96] LIGHTSTONE F.C., BRUICE T.C., "Ground state conformations and entropic and enthalpic factors in the efficiency of intramolecular and enzymatic reactions. 1. Cyclic anhydride formation by substituted glutarates, succinate, and 3,6-endoxo-Δ4-tetrahydrophthalate monophenyl esters", *Journal of the American Chemical Society*, vol. 118, pp. 2595–2605, 1996.

[MIL 87] MILLER S.L. "Which organic compounds could have occurred on the prebiotic earth?", *Cold Spring Harbor Symposia on Quantitative Biology*, vol. LII, pp. 17–27, 1987.

[OGL 01] OGLE J.M., BRODERSEN D.E., CLEMONS JR. W.M. *et al.*, "Recognition of cognate transfer RNA by the 30S ribosomal subunit", *Science*, vol. 292, pp. 897–902, 2001.

[REE 81] REEVES H., *Patience dans l'azur - L'évolution cosmique*, Éditions du Seuil, Paris, 1981.

[ROZ 16] ROZOV A., WESTHOF E., YUSUPOV M. *et al.*, "The ribosome prohibits the G•U wobble geometry at the first position of the codon-anticodon helix", *Nucleic Acids Research*, vol. 44, pp. 6434–6441, 2016.

[TUR 87] TURNER D.H., SUGIMOTO N., JAEGER J.A. *et al.*, "Improved parameters for prediction of RNA structure", *Cold Spring Harbor Symposia on Quantitative Biology*, vol. 52, pp. 123–133, 1987.

[VAN 16] VAN DER GULIK P.T., HOFF W.D., "Anticodon modifications in the tRNA set of LUCA and the fundamental regularity in the standard genetic code", *PLoS One*, vol. 11, p. e0158342, 2016.

[VEN 08] VENDEIX F.A., DZIERGOWSKA A., GUSTILO E.M. *et al.*, "Anticodon domain modifications contribute order to tRNA for ribosome-mediated codon binding", *Biochemistry*, vol. 47, pp. 6117–6129, 2008.

[WAN 17] WANG J., LEHMANN J., "Commaless Code", in ROITBERG B. (ed.), *Reference Module in Life Sciences*, Elsevier, Science Direct, 2017.

[WOE 66] WOESE C.R., DUGRE D.H., SAXINGER W.C. *et al.*, "The molecular basis for the genetic code", *Proceedings of the National Academy of Sciences of the United States of America*, vol. 55, pp. 966–974, 1966.

[YAN 13] YAN H., HUANG W., YAN C. *et al.*, "Structure and mechanism of a nitrate transporter", *Cell Rep*, vol. 3, pp. 716–723, 2013.

[ZEN 14] ZENG X., CHUGH J., CASIANO-NEGRONI A. *et al.*, "Flipping of the Ribosomal A-Site Adenines Provides a Basis for tRNA Selection", *Journal of Molecular Biology*, vol. 426, pp. 3201–3213, 2014.

Telomeres and Telomerases: Structural Diversity for the Same Role

7.1. Introduction

The first cells with linear chromosomes appeared approximately 1 billion years ago. The extremities of these linear chromosomes were called telomeres, derived from the Greek words *telos* (end) and *mere* (part), namely the end of chromosomes. The presence of chromosome extremities results in two main problems to the cell: the problem of protecting chromosome extremities [DEL 09] and the problem of replicating these regions [MAE 17].

In the cell, there are repair systems that detect and repair DNA double-strand breaks (considered as damage to DNA), by allowing the fusion of the broken DNA extremities to restore an intact chromosome [MLA 16]. It is important that the natural chromosome extremities are not identified as extremities derived from double break, as it would risk causing the fusion of chromosomes and then a chromosomal instability. Similarly, these extremities should not be the target of exonucleases, which could result in the degradation of chromosomes. H.J. Muller, in the 1930s, observed that deletions or inversions induced by X-ray radiation of the whole genome were nearly undetectable in the terminal regions of Drosophila chromosomes [MUL 38]. He concluded that the chromosome extremities must have a specific structure capable of stabilizing chromosomes. At the same time,

Chapter written by Carole SAINTOMÉ.

B. McClintock demonstrated that corn chromosomes which suffered double-strand breaks were capable of fusing together, unlike their natural extremities, which remained stable [MCC 39]. Following these results, the authors defined telomere as a specific structure likely to protect chromosomes from the potential fusions of their extremities.

In the 1970s, the end replication problem was raised by J.D. Watson: the enzymatic complex of polymerase DNA is incapable of synthesizing the extremity of linear DNA molecules; telomeric DNA is then the object of a shortening [WAT 72]. A replication mechanism was described to explain telomere shortening at the end of each cell division. In fact, the progression of the DNA polymerase is always carried out in the 5'-3' direction, which explains that the two-strand replication progress of the replication fork is not identical. It is continuous for the leading strand and discontinuous for the lagging strand. In this last case, small fragments (Okazaki fragments), with a size of approximately 100 nucleotides in eukaryotes, containing the RNA primer and the DNA fragment, are synthesized. RNA primers are then excised and replaced by DNA fragments synthesized by the DNA polymerase I, which is then joined together by a ligase. All the gaps formed by removing RNA primers are filled with DNA, except the gap formed at the 5' extremity of the telomere-lagging stand (strand C). Consequently, the strand neo-synthesized from the lagging strand will be shorter than the original strand. This phenomenon was called the end replication problem. Furthermore, extremity resection mechanisms after replication also contribute to the shortening of this region. Thus, at each cell division, telomeres are shortened. The relationship between the length of the telomere and the aging was highlighted by A. Olovnikov, who offered a hypothesis linking cell senescence to the replication problem [OLO 73]. He explained the limited cell proliferation (Hayflick limit) by the erosion of telomeres at each cell division. Thus, telomeric shortening corresponds to an internal aging clock. However, some stem and germinal cells have the capacity of indefinite division: the presence of a telomerase activity, which maintains telomere length, is the origin of this unlimited division.

To understand how the cell manages these two problems of protection and replication of extremities, numerous studies were undertaken in order to first determine the nature of telomeres, which are specific nucleoprotein structures formed by the association of telomeric DNA and specific proteins.

7.2. Nature of chromosome extremities

Telomeres have a significant structural diversity according to the organisms [FUL 14]. In eukaryotes, the structure is very conserved (Figure 7.1(a)). It is composed of a double-strand region and a single-strand extension on the 3' side, called G-overhang. The length of the double-strand region and the single-strand region varies according to species, the type of cell and the different chromosomes. In humans, the length of telomeres varies between 2 and 15 kb for the double-strand region and between 150 and 300 bases for the single-strand region. In some viruses and bacteria with linear chromosomes and in mitochondria, very varied structures are found, with a double-strand region terminated with either a 3' or 5' single-strand region, with a covalently bound hairpin structure or with a covalently bound protein.

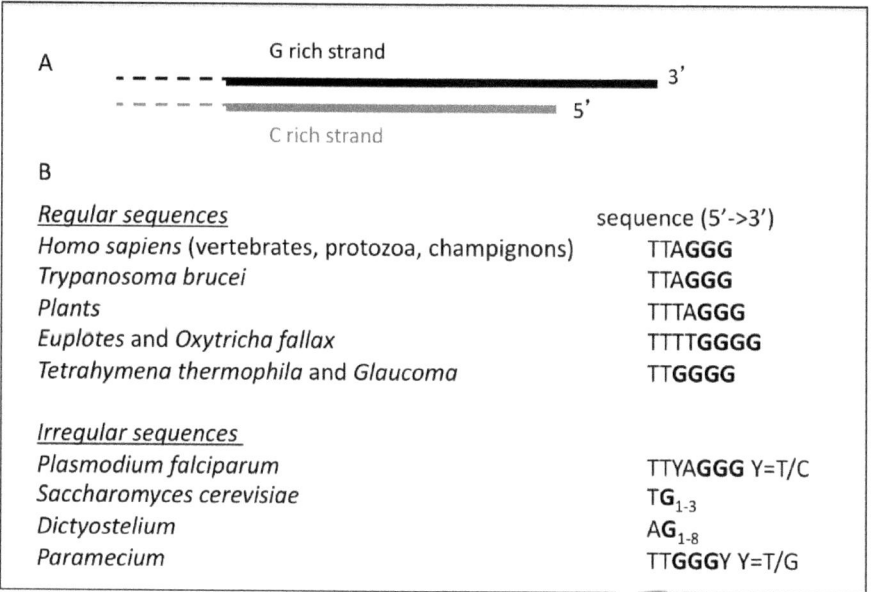

Figure 7.1. *Structure (a) and telomeric sequences (b) of various eukaryotic organisms*

In most eukaryotic cells, telomeric DNA consists of short tandem repeats, a non-coding sequence, which varies from one species to the other. This sequence is most often rich in guanine on the 5' to 3'-oriented strand toward the chromosome extremity, but sequences rich in G+T (Figure 7.1(b)) are also found. The 5'TTAGGG3' sequence represents the chromosome extremity of vertebrates; other telomeric sequences were identified in protozoa, yeasts and higher plants (Figure 7.1(b)). The first telomere TTGGGG motif was characterized in the ciliate *Tetrahymena thermophila* by E. Blackburn [BLA 78]. The telomeric DNA sequencing in different species shows that the DNA sequence at chromosome extremities is relatively well conserved (TxAyGz type consensus sequence); the TTAGGG sequence represents the canonical sequence of most eukaryotes. As with eukaryotes, in some viruses and bacteria and in mitochondria, telomeric DNA can consist of short tandem repeats or a long repetition of inverted sequences called TIR (Telomere Inverted Repeat). In drosophila, transposable elements form telomeres (except in eukaryotes, which we do not discuss in this chapter).

7.3. Telomeres in eukaryotes

Numerous structural studies of eukaryotic telomeric sequences showed that the latter, because of their richness in guanines and their tandem repetition, could form specific structures, the presence of which at chromosome extremities helped to avoid telomere degradation and telomeric fusions. These structures then played a protecting role.

7.3.1. *Structures of sequences rich in G: G-quadruplexes*

Numerous *in vitro* studies showed that sequences rich in guanines form specific structures, called G-quadruplexes (G4), which are very stable under physicochemical conditions [RHO 15]. These parallel or antiparallel structures (Figures 7.2(a) and (b)) are formed following the hydrophobic stacking of at least two G-quartets (Figure 7.2(c)), intercalating a monovalent cation in the central cavity. The cation is chelated by the four oxygens of the guanine carbonyl functions. The number of G-quartets is equivalent to the number of adjacent guanines in the primary DNA sequence including G repetitions. The telomeric sequence of vertebrates has a significant polymorphism noted by *in vitro* biophysical studies.

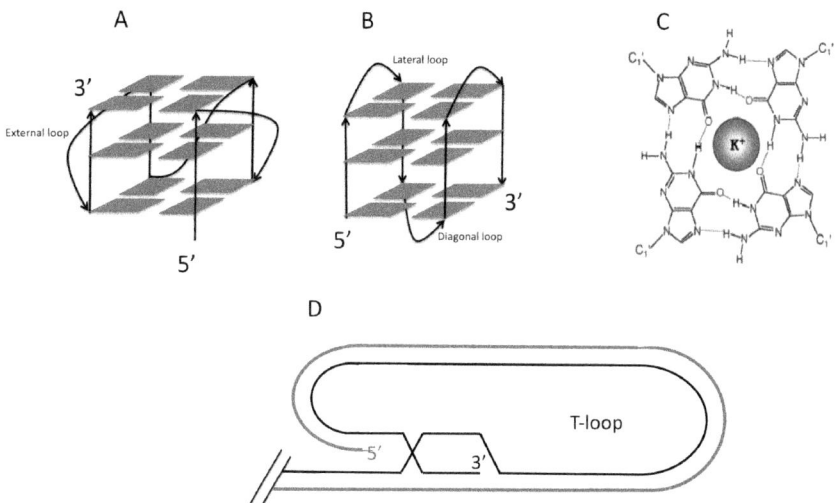

Figure 7.2. *Structures of the human telomeric sequence: parallel a), antiparallel b), G-quadruplex, G-quartet c) and t-loop d).*

The first evidence of the existence of G4s *in vivo* comes from the work carried out by A. Plückthün and coworkers on ciliates, whose particularity is to possess two nuclei: one reproductive diploid micronucleus and one macronucleus used for the synthesis of proteins and which contains millions of minichromosomes. The use of an affined synthetic fluorescent antibody specific to a parallel G4 structure, which is formed by the telomeric sequence of the ciliate *Stylonychia lemnae*, showed a pronounced marking in the isolated macronuclei of ciliates [SCH 01]. Very recently, telomeric G4 structures have been highlighted by means of the observation of a labeling of chromosome extremities, in human normal and tumor cells, after treatment with a specific tritiated G4 ligand (3H-360A) [GRA 05]. Numerous studies on the interactions of G4 ligands and G4 structures, as well as on the effect of G4 ligands in cells, provided significant evidence of the presence of G4s *in vivo*. Furthermore, a bioinformatic study showed that the human genome contains approximately 700,000 sequences likely to form G4 structures called PQS (Putative G-quadruplex Sequences). However, we do not know how many of them form them *in vivo*. Sequences are not randomly distributed in the genome, because they are mainly found in telomeres and in sequences of oncogenes or proto-oncogenes. The existence of these sequences in these key regions of the eukaryotic genome suggests that G4s could play an important role *in vivo* [RHO 15].

7.3.2. t-loop

Electronic microscopy studies on purified telomeric extracts of human and mice cells made it possible to visualize a t-loop structure (for telomeric-loop) [GRI 99]. Such a conformation was also highlighted in *Trypanosoma brucei*, the pea, ciliates and in chicken and mice cells. In a similar manner to the first stages of the homologous recombination, this large loop would result from the invasion of the 3' extremity in the telomeric double-strand region (Figure 7.2(d)). The G-strand of the double-stranded DNA, moved by the 3' extremity, forms the D-loop (Displacement t-loop). The formation of this structure would be possible for a relatively long telomeric single strand thanks to the association of specific proteins. It was suggested that at least 100 nucleotides of the single-strand extremity contribute to the formation of this t-loop. The loop size can vary, as 1 kb as well as 25 kb t-loops were observed in human cells. *In vitro*, the formation of the t-loop is mediated by the TRF2 telomeric protein, one of the proteins of the shelterin complex. However, *in vivo*, it seems that TRF2 requires the intervention of other proteins to generate the t-loop.

These specific structures could contribute to the protection of chromosome extremities. However, it was shown that their presence is an obstacle to the progression of the replication fork and can lead, if they are not eliminated, to a telomere shortening. In order to avoid this, numerous proteins intervene to solve these structures.

7.3.3. Proteins at telomeres

Like the rest of the chromosome, telomeric DNA is associated with proteins whose nature is specific to this region. Their role is to protect chromosome extremities by stabilizing specific structures, such as the t-loop. The loss of these proteins results in a telomeric instability. These telomeric proteins are very diversified according to the organisms. The proteins specific to human telomeres form a complex called shelterin or "telosome" (Figure 7.3(a)). The shelterin is approximately 1 MDa; it is composed of six proteins [DEL 05]. It specifically recognizes TTAGGG telomeric repetitions thanks three proteins: TRF1 and TRF2 (telomeric repeat binding factors 1 and 2), which bind to the telomeric double-strand part, and POT1 (protection of telomeres 1), which is associated with the TTAGGG single-strand repetitions of the 3' extremity. The other proteins of the shelterin complex,

TIN2 (TRF2- and TRF1-interacting nuclear protein 2), Rap1 (repressor activator protein 1) and TPP1 (TINT1/PTOP/PIP1 protein), are used as adaptors between TRF1, TRF2 and POT1. In the yeast *S. pombe*, proteins similar to humans' are found (Figure 7.3(b)), while they can be very different, such as in the yeast *S. cerevisiae* (Figure 7.3(c)).

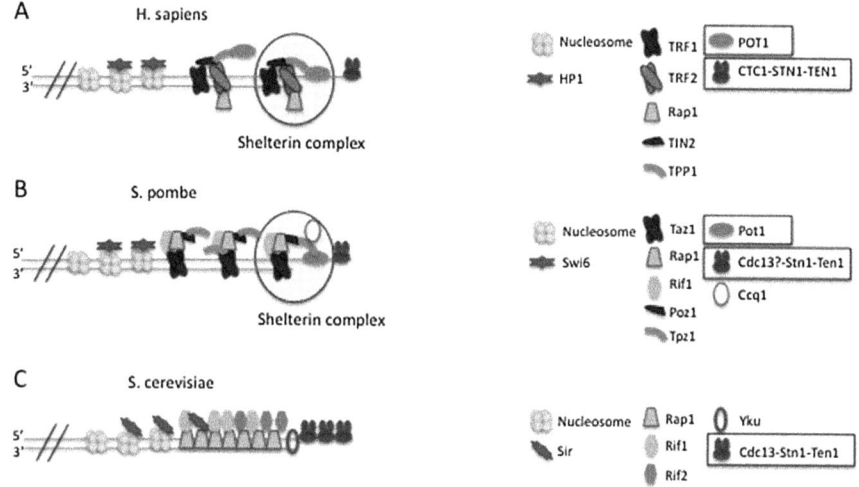

Figure 7.3. *Telomeric proteins. The proteins with OB-fold motif are in boxes. For a color version of this figure, see www.iste.co.uk/grandcolas/biodiversity.zip*

Among all the proteins forming the shelterin of these different organisms, those specifically binding the telomeric single-strand DNA have at least one binding motif called OB-fold (meaning oligosaccharide–oligonucleotide binding domain). The conservation of this OB-fold motif during evolution suggests the significant role of this motif in the specific recognition of telomeric sequences at the G-overhang [HOR 11]. In addition, the structural similarities of these proteins with the Replication Protein A (protein very conserved in eukaryotes with no sequence specificity) suggest that they are derived from an RPA-like protein. These data raise the issue of the nature of the factors that guided the evolution of a protein with no sequence specificity, such as RPA toward specific proteins of the telomeric DNA sequence, such as POT1, as well as the co-evolution between the OB-fold proteins and their telomeric sequences.

Finally, the presence of specific telomeric DNA structures and telomeric proteins converges toward the same role of protecting telomeres. However, when repeated sequences become too short, telomeres are no longer able to play their protecting role. Chromosome extremities are then recognized as double-strand DNA breaks, and molecular monitoring mechanisms are activated (checkpoint), a function of which is to stop the cell cycle. In this context, cells enter in senescence and no longer divide, in order to prevent chromosomal events that would cause a chromosomal instability.

7.4. Maintenance of telomeres by telomerase in eukaryotes

The telomere replication problem was understood thanks to the discovery of telomerase in 1985, resulting from the work of Elizabeth Blackburn and her collaborators, who received the Nobel Prize for Medicine in 2009. Telomerase was identified for the first time in *Tetrahymena thermophila* when Carol Greider and Elizabeth Blackburn noted that the telomeric sequence was extended *in vitro* by identical sequences with telomere repetitive motif [GRE 85]. Telomerase activity was then identified in numerous other organisms: by means of an *in vitro* elongation assay with cell extracts, a telomerase activity was identified in *Oxytricha*, Euplotes and yeast; human telomerase was identified in 1989, first in HeLa tumor cells and then in leucocytes derived from healthy persons and patients suffering from leukemia.

Telomerase is a protein complex associated with an RNA sequence essential for the catalytic activity. The telomerase enzyme activity *in vitro* is carried by the TERT (telomerase reverse transcriptase) protein and the TER RNA (telomerase RNA). Numerous *in vivo* studies showed the interaction of TERT with other proteins, which could be essential for the telomerase enzyme activity. Other telomerase protein components were identified thanks to biochemical and genetic studies. Some were directly involved in telomere replication, and others in telomerase regulation. Most of these additional subunits play a part in telomerase assembly, conformation and nuclear location [SCH 15, NAD 13].

7.4.1. *Structure of the catalytic subunit: very conserved domain of the RT*

The catalytic part of human telomerase called hTERT was identified in 1997 [NAK 97]. hTERT contains the seven conserved motifs (1, 2, A, B', C, D and E) of the HIV reverse transcriptase in its central region called RT (Figure 7.4(a)). Like other polymerases, the TERT protein contains a very conserved triad of aspartic acids, the mutation of which destroys the enzyme catalytic activity. The hTERT N-terminal region contributes to the unique properties of the telomerase, such as its association with its telomeric DNA (NTE domain) and TER RNA (TRBD domain) (Figure 7.4(a)), the fixation of other protein components and the modulation of its processivity.

7.4.2. *RT stretching mechanism*

C. Greider and E. Blackburn were the first to offer a model to maintain the telomere length through telomerase. In this model, telomerase uses its TER RNA as a template to add nucleotides to the chromosome extremity thanks to the reverse transcriptase activity present in the catalytic subunit (TERT).

The mechanism includes the following three stages (Figure 7.4(b)):

I – Telomere–telomerase association and binding of the RNA template to the 3' single-strand extremity of the telomeric DNA through sequence complementarity;

II – Elongation through reverse transcription of the RNA template and addition of nucleotides to the 3' extremity of the telomeric DNA;

III – Translocation stage: once the reverse transcription of the RNA template sequence is finished, a translocation of the telomerase complex occurs to facilitate another synthesis cycle.

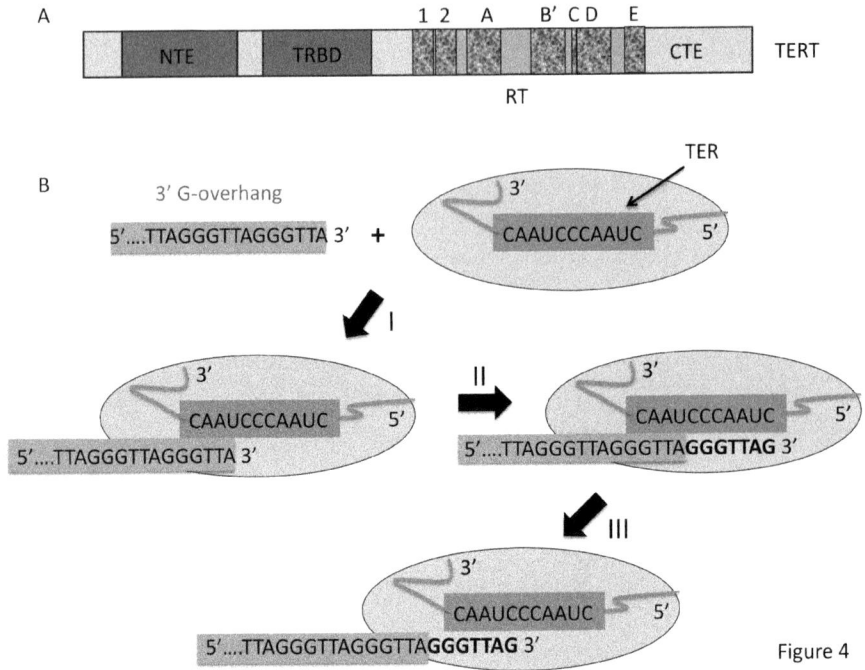

Figure 4

Figure 7.4. *Telomerase structure and action mechanism. (a) Different domains of the telomerase TERT catalytic subunit. (b) Different elongation stages of the 3' G-overhang by telomerase. For a color version of this figure, see www.iste.co.uk/grandcolas/biodiversity.zip*

7.4.3. RNA subunit structure

C Greider and E. Blackburn found the existence of an RNA sequence that would help to ensure the elongation of telomeric repetitions in *Tetrahymena*. This hypothesis came from the fact that the treatment with an RNAse inactivated the synthesis of telomeric repetitions *in vitro* [GRE 87]. Later, this hypothesis was validated with an experiment cloning an RNA coding gene in *Tetrahymena*. The transcription of this gene produced an RNA with a portion of its sequence that was CAACCCAA, which is a sequence complementary to the repetitive motif of the telomeric DNA in the same organism.

The RNA part was cloned in several different organisms and shows variability in length and sequence, as well as in the general structure. For example, the TER RNA size can vary between 100 (*C. elegans*) and 1,000 (*S. cerevisiae*) nucleotides. In humans, TER has 451 nucleotides. If their sequence and structure are very variable, we find, however, conserved domains in all species (Figure 7.5(a)): a template sequence (Template), complementary to telomeric sequences; a TBE (Template Boundary Sequence) hairpin motif, which helps to prevent DNA synthesis outside the template region, and a transactivating domain, called CR4-CR5 [POD 16a]. These domains were conserved during evolution, because they played an essential role for telomerase activity. In most known TRs, a pseudoknot structure is present and located near the template; even if the size and form of this pseudoknot vary according to species, its presence is crucial for telomerase activity. However, in some organisms, the pseudoknot does not seem essential for telomerase activity (in the ciliate *Tetrahymena*) or does not even exist (in *Flagella Trypanosoma*). The absence of the pseudoknot in basal eukaryotic species and its minimal role in the telomerase activity of ciliates suggest that the pseudoknot appeared during evolution in more evolved eukaryotes [POD 16b]. Thus, the TR common ancestor should not have any pseudoknot in the template region.

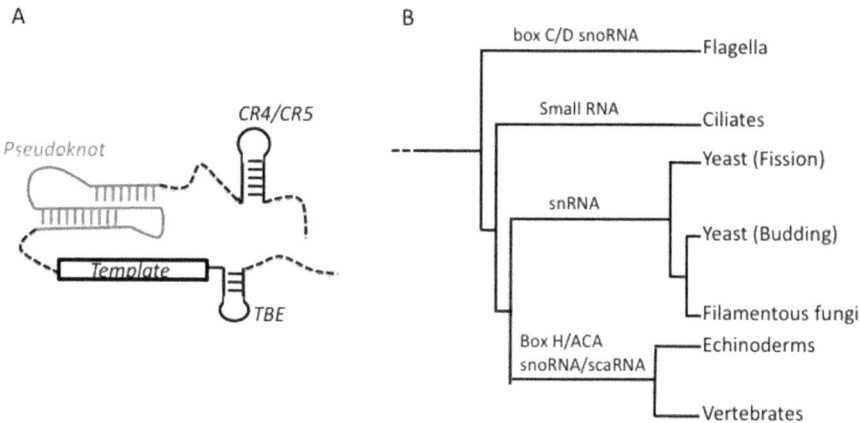

Figure 7.5. *TER RNA. (a) Conserved motifs of eukaryotic TRs. (b) Different biogenesis pathways of the telomerase ribonucleoprotein complex.*

If TR has relatively well-conserved domains during evolution, contributing to the telomerase catalytic activity, then it also has other structural domains; however, they are very diversified, mainly located on the TR 3' side and contribute to the telomerase biogenesis [POD 16b]. These domains are involved in very distinct RNA metabolic pathways making post-transcriptional modifications and RNA splicing possible, as well as the nuclear assembly, stabilization and location of the telomerase complex; to do so, they form binding sites for the various proteins involved in these different biogenesis pathways (Figure 7.5(b)). In vertebrates, TR has the H/ACA motif, which is specific to the biogenesis pathway through snoRNAs (small nucleolar RNAs) and scaRNAs (small Cajal RNAs), as well as a binding motif (CAB) of the TCAB1 protein, involved in the specific location in Cajal bodies (nuclear compartment where post-transcriptional modifications and RNA splicing occur). Echinoderms take the biogenesis pathway of snoRNAs and scaRNAs, as they also contain the H/ACA motif. In yeasts, it is the biogenesis pathway through snRNAs (small nuclear RNAs) that is taken, however with diversified maturation mechanisms according to species (intervention of the Nrd1–Nab3–Sen1 complex or the spliceosome). In ciliates, biogenesis takes the biogenesis pathway of the small RNA transcripts of the RNA polymerase III. In flagella, it is the biogenesis pathway of snoRNAs, as they also contain the C/D motif.

Although the TRs of the various species are very different in size, sequence and structural motifs, they all have an interaction domain with TERT, corresponding to the domain involved in the telomerase catalytic activity. They also have an interaction domain with numerous and various proteins, which are involved in the biogenesis of the ribonucleoprotein complex. The interaction domain with TERT is well conserved compared to the domain involved in biogenesis, which radically varies from one eukaryotic lineage to the other. Thus, during evolution, TER maintained the motifs that were essential for interacting with the highly conserved telomerase, while assimilating new and various recognition motifs for the various proteins involved in the telomerase biogenesis.

To conclude, telomeres are nucleoprotein structures with a variable nature in terms of DNA size and sequence, as well as proteins present. In eukaryotes, the telomeric sequence is a tandem repeated sequence rich in guanines and capable of forming G-quadruplex or t-loop structures. This structural characteristic is conserved in numerous organisms and could be a factor essential for telomere protection. Finally, the structural and functional study of the telomerase involved in telomere maintenance shows that, if the catalytic subunit (TERT) was conserved during evolution, the RNA subunit (TER) very quickly evolved from a minimal structure containing the motifs essential for telomerase activity. The appearance during evolution of new structural motifs would have helped diversification for the telomerase biogenesis.

7.5. Bibliography

[BLA 78] BLACKBURN E.H., GALL J.G., "A tandemly repeated sequence at the termini of the extrachromosomal ribosomal RNA genes in Tetrahymena", *Journal of Molecular Biology*, vol. 120, pp. 33–53, 1978.

[DEL 05] DE LANGE T., "Shelterin: the protein complex that shapes and safeguards human telomeres", *Genes and Development*, vol. 19, pp. 2100–2110, 2005.

[DEL 09] DE LANGE T., "How telomeres solve the end-protection problem", *Science*, vol. 326, p. 948, 2009.

[FUL 14] FULCHER N., DERBOVEN E., VALUCHOVA S. *et al.*, "If the cap fits, wear it: an overview of telomeric structures over evolution", *Cellular and Molecular Life Sciences*, vol. 71, pp. 847–865, 2014.

[GRA 05] GRANOTIER C., PENNARUM G., RIOU L. *et al.*, "Preferential binding of a G-quadruplex ligand to human chromosome ends", *Nucleic Acids Research*, vol. 33, pp. 4182–4190, 2005.

[GRE 85] GREIDER C., BLACKBURN E.H., "Identification of a specific telomere terminal transferase activity in Tetrahymena extracts", *Cell*, vol. 43, pp. 405–413, 1985.

[GRE 87] GREIDER C., BLACKBURN E.H., "The telomere terminal transferase of Tetrahymena is a ribonucleoprotein enzyme with two kinds of primer specificity", *Cell*, vol. 51, pp. 887–898, 1987.

[GRI 99] GRIFFITH J.D., COMEAU L., ROSENFIELD S. *et al.*, " Mammalian telomeres end in a large duplex loop", *Cell*, vol. 97, pp. 503–514, 1999.

[HOR 11] HORVATH M.P., "Structural anatomy of telomere OB proteins", *Critical Reviews in Biochemistry and Molecular Biology*, vol. 46, pp. 409–435, 2011.

[MAE 17] MAESTRONI L., MATMATI S., COULON S., "Solving the telomere replication problem", *Genes*, vol. 8, 2017.

[MCC 39] McCLINTOCK B., "The behavior in successive nuclear divisions of a chromosome broken at meiosis", *Genetics*, vol. 25, pp. 405–416, 1939.

[MLA 16] MLADENOV E., MAGIN S., SONI A. *et al.*, "DNA double-strand-break repair in higher eukaryotes and its role in genomic instability and cancer: cell cycle and proliferation-dependent regulation", *Semin. Cancer Biol.*, vols 37–38, pp. 51–64, 2016.

[MUL 38] MULLER H.J., "The remarking of chromosomes", *Collecting Net.*, vol. 13, p. 5837777, 1938.

[NAK 97] NAKAMURA T.M., MORIN G.B., CHAPMAN K.B. *et al.*, "Telomerase catalytic subunit homologues from fission yeast and human", *Science*, vol. 277, pp. 955–959, 1997.

[NAN 13] NANDAKUMAR J., CECH T.R., "Finding the end: recruitment of telomerase to the telomere", *Nature Reviews Molecular Cell Biology*, vol. 14, pp. 69–82, 2013.

[OLO 73] OLOVNIKOV A.M., "The incomplete copying of Template margin in enzymic synthesis of polynucleotides and biological significance of the phenomenon", *Journal of Theoretical Biology. A Theroy of Marginotomy*, vol. 41, pp. 181–190, 1973.

[POD 16a] PODLEVSKY J.D., LI Y., CHEN J.L., "The functional requirement of two structural domains within telomerase RNA emerged early in eukaryotes", *Nucleic Acids Research*, vol. 44, pp. 9891–9901, 2016.

[POD 16b] PODLEVSKY J.D., CHEN J.L., "Evolutionary perspectives of telomerase RNA structure and function", *RNA Biology*, vol. 13, pp. 720–732, 2016.

[RHO 15] RHODES D., LIPPS H.J., "G-quadruplexes and their regulatory roles in biology", *Nucleic Acids Research*, vol. 43, pp. 8627–8637, 2015.

[SCH 01] SCHAFFITZEL C., BERGER I., POSTBERG J. *et al.*, "In vitro generated antibodies specific for telomeric quanine-quadruplex DNA react with Stylonychia lemnae macronuclei", *Proceedings of the National Academy of Sciences*, vol. 98, pp. 8572–8577, 2001.

[SCH 15] SCHMIDT J.C., CECH T.R., "Human telomerase: biogenesis, trafficking, recruitment and activation", *Genes and Development*, vol. 29, pp. 1095–105, 2015.

[WAT 72] WATSON J.D., "New biology. Origin of concatemeric T7 DNA?", *Nature*, vol. 239, pp. 197–201, 1972.

8

Globalization and Infectious Diseases

8.1. Introduction

In this chapter, we will discuss the demographic and geographic dynamic of a human pathogen, *Mycobacterium tuberculosis*, which is the causative agent of tuberculosis. Constructing a model of the evolution of this bacterium requires multidisciplinary expertise on host–pathogen coevolution, the history of human migration, the Neolithic revolution and the history of domestication.

Since the mid-20th Century, the success of *M. tuberculosis* has not only heavily relied on the demographic explosion of its main reservoir *Homo sapiens*, but also on its growing resistance to antibiotics and the rise of compensatory mutations. In order to highlight a few common patterns shared by different bacteria, as well as the universality of the evolutionary and adaptive forces at stake, we will also briefly discuss some other bacterial pathogens, namely *Staphylococcus aureus* and *Salmonella typhi*.

Koch's bacillus is a scourge which has been affecting human societies for thousands of years, reaching its climax at the end of the 19th Century. This relative temporal proximity implies that many people around us had

Chapter written by Thierry WIRTH.

grandparents or great-grandparents who died from pulmonary infection complications. Phthisis is a disease which resulted in paleness, glassy eyes, lethargy and weight loss. This gave patients a certain charm. Numerous artworks of that era vividly depicted this romantic imagery. Famous examples include Mimi in *La Bohème* by Puccini, the courtesan Marguerite Gautier in *The Lady of the Camellias* by Alexandre Dumas, fils, Hans Castorp in *The Magic Mountain* by Thomas Mann, and *The Sick Child* by the famous Norwegian painter Edvard Munch portraying the very last moments of his dying sister. Despite its long history, tuberculosis is far from a "disease of the past". Rather it remains a major threat to the contemporary world. Even though better hygiene conditions and the introduction of antibiotics in the 1950s have significantly slowed down its spread, an estimated 2 billion people on the planet are still healthy carriers, i.e. they have contracted the germs but display no symptoms, because their immune system is able to control the pathogen without falling ill. However, nearly 10% of this population will finally develop active TB in various forms (which are mostly pulmonary, but it can also attack bones, urinary tracts and lymph nodes) [GEN 12]. According to the World Health Organization, an estimated 10.4 million people became ill with TB in 2016, and it caused the death of almost 1.7 million people in the same year, making it the ninth leading cause of death in the world and the infectious agent that caused the most deaths, ranking above HIV/AIDS [WOR 17].

Geographically, tuberculosis is relatively more prevalent in countries with high HIV/AIDS prevalence rates, such as South Africa and its neighboring countries, as the immunodeficiency caused by HIV very often leads to acute TB. In 2016, there were 476,774 reported cases of HIV-positive tuberculosis, i.e. 46% of all estimated incidents [WOR 17]. Another main geographic area (in absolute number of cases) is continental countries with high population density and growth, such as China and India.

Thanks to national and international efforts, the mortality rate of tuberculosis has dropped by 50% since 1990 [WOR 17]. However, this progress is hindered by the emergence of bacterial strains that are resistant to antibiotics. Based on the types of antibiotics they are resistant to, they are identified as multidrug-resistant (MDR), extensively drug-resistant (XDR) and even totally drug-resistant (TDR). The latter brings us back to a pre-antibiotics situation where care was only palliative, as no antibiotic cocktail has proven effective against these germs. The *Mycobacterium tuberculosis* MDR and XDR strains mostly come from the former Soviet Republics, while numerous cases have also been found in China and India. Treatment failure rates among patients with MDR strains are up to 50% in these regions, and we believe that nearly 90% of all patients do not benefit from sufficient treatment. The high cost of MDR treatment is also an important hurdle. It is about 134,000 dollars in the United States [MAR 14] and approximately 200,000 euro in Europe. The intimidating prices fostered so-called "health migration". This is a phenomenon of people going abroad to seek therapeutic assistance that they are unable to obtain in their country of origin. This situation is further facilitated by the mandatory declaration of TB and the treatment order in numerous EEC countries.

8.2. Origins of tuberculosis

Questions that have long tormented humanity, such as "who are we?", "where do we come from?" and "where are we going?", can be transposed to any living organism. What about the origin of tuberculosis, where does this pathogen come from? Some answers are derived from direct observations of bone lesions caused by *M. tuberculosis*. Paleopathologists have discovered tuberculosis signatures on Egyptian mummies which are more than 3,000 years old, by using molecular typing techniques to amplify DNA traces [ZIN 03]. Typical lesions of Pott disease were found in tombs, dating from the beginning of the Copper Age [POS 15]. Although it has been considered as a zoonosis for long time, tuberculosis was in fact not transmitted, as often indicated in scientific literature, from bovines to humans. In reality, *M. tuberculosis* is a sub-species complex or a group of evolutionary lineages with multiple origins and different hosts (Figure 8.1). Depending on the type of their hosts, we can divide them into two independent clades: 1) strictly human strains, including modern and ancestral ones, and 2) animal strains which attack rodents, pinnipeds, bovines, caprines, meerkats and oryx. With the help of new sequencing technologies, scientists have found that strains in non-human mammals have smaller genomes with scars reflecting secondary gene losses that are not essential to survive in their new hosts (for example, those specific to the immune response of *Homo sapiens*). These characteristics demonstrate the derived nature of these strains and the evolutionary precedence of typically human strains.

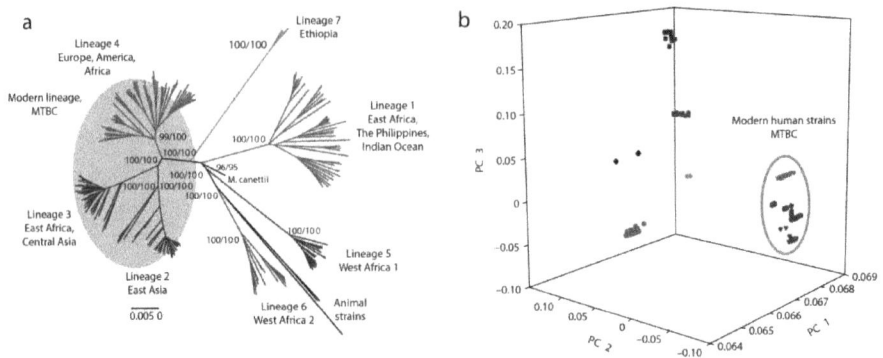

Figure 8.1. *a) Phylogeny of the M. tuberculosis complex from 220 strains representing the overall diversity. b) Principal component analysis relying on 34,167 detected single-nucleotide polymorphisms. For a color version of this figure, see www.iste.co.uk/grandcolas/biodiversity.zip*

Further observation of the phylogeny of the *M. tuberculosis* complex (MTBC) revealed that the main human lineages have remarkable biogeographic imprints [COM 13]. For example, some lineages are strictly Asian; some others African, such as *M. africanum*, and some are typical of European and North-American populations. In fact, based on the intimate host–pathogen relationship and co-evolutionary consequences, the team of Sébastien Gagneux managed to uncover a coevolution of the phylogenic tree of *M. tuberculosis* and that of human mitochondrial genes [COM 13]. Strikingly, both trees have the same African origin, close topologies and identical geographical clades. Based on the comparison of the *M. tuberculosis* lineage of Southeast Asia with that of the macro-haplogroup M of the human mitochondria, the authors succeeded in estimating tuberculosis' first appearance to be slightly posterior to the out-of-Africa migration, namely approximately 65,000 years ago. Even if the dating process might seem a bit tautological, it is nevertheless considered as the best estimation to date. It provides us with a realistic scenario of the emergence and spatial diffusion of the disease during the Neolithic expansion.

8.3. Emergence of MDR-Beijing lineage

We were able to discover tuberculosis' historical origins, thanks to phylogenetic and population genetic tools. Studies of strains derived from the Beijing lineage using similar tools allowed us to further understand the contemporary worldwide success of this major pathogen. The so-called

Beijing strains came to the attention of medical professionals in the 2000s due to their resistance to all first-generation antibiotics, and even to second-generation ones. Their propagation immediately became a non-negligible health issue. In France, in 2012, 100% of XDR strains were Beijing strains that came from immigrants from the former Soviet Republics [BER 13]. This worrisome situation was also found in Abkhazia, where nearly one third of all new MDR tuberculosis cases were caused by Beijing strains. In order to better understand the success of this lineage and its propagation, a consortium of researchers from the Tuberculosis Reference Center in Germany, the *Institut Pasteur* of Lille, the American CDC and the *Muséum National d'Histoire Naturelle* of Paris analyzed almost 5,000 Beijing strains collected in 100 different countries [MER 15]. By using mycobacterial interspersed repetitive units (MIRUs) as mini satellite-like genetic markers, the team revealed once again a strong biogeographical character of Beijing strains. Several clonal complexes (CC numbered from 1 to 7) were identified, among which were two particularly epidemic clones from Central Asia (CC1) and Russia (CC2). The allelic diversity was the highest in CC6, and in an ancestral lineage named BL7. By analyzing the distribution of clonal complexes worldwide, the authors found that the greatest genetic diversity of Beijing *M. tuberculosis* was in East Asia, and that the most ancestral strains came from Korea and Japan, which spread very little outside Asia. Gradients or clines are also identifiable on the global map. We can see the so-called Pacific lineage spreading outside Asia with an increasing frequency along an east–west axis, while the Russian clone mostly remained in Central and Eastern Europe. In numerous evolutionary settings, finding the area with the greatest genetic diversity of a species often signifies the discovery of its birthplace (see *H. sapiens* in Africa). By applying the same approach, the researchers demonstrated that the Beijing strain's genetic diversity eroded when moving away from the Yangtze River Basin. This negative correlation is stronger at the Eurasian continent level ($r^2 = 0.63$) than at the global level ($r^2 = 0.22$). This gap can be quite easily explained by the so-called Boeing effect (long-distance migration facilitated by airplanes), as the most deviating sampling points correspond to regions with significant intercontinental immigration waves (e.g. South Africa). A heat map analysis made it possible to confirm these results. It also suggested that the reasons behind the diversity of Beijing strains in contemporary Africa goes beyond simple natural diffusions. In fact, this diversity excess could be explained by the gradual replacement of Françafrique by Chinafrica. In recent years, the increasing economic interests shared between the oriental giant and its African partners are well reflected by the presence of nearly two million Chinese workers and engineers on the African continent, who have probably brought those strains to their new country of residence. This working hypothesis, although requiring a more exhaustive sampling before drawing final conclusions, enables us to grasp the impact of macroeconomics and international migration flows on the global epidemics map.

By analyzing MIRUs' profiles in a succession of temporal isolates collected from patients, we estimated the mutation rate of these MIRUs to be 10^{-4} change/locus/year. This implies that the common ancestor of Beijing lineage strains is nearly 7,000 years old. This dating is not unrelated to a previously presented scenario, in which Thierry Wirth and his colleagues determined that the main secondary radiations of human tuberculosis strains must have occurred in Mesopotamia during domestication [WIR 08]. This synergistic effect seems to have occurred also in China, where rice cultivation and pig farming are 7,000–8,000 years old. The selective advantage of an epidemic pathogen in a domestication context is easy to understand; adequate food from agriculture boosted population growth causing the transmission chain to extend and to densify, making the situation completely different from the hunter-gatherer era.

In addition, the molecular tool can also be applied to smaller time scales. Unlike the common belief that the acquisition of resistance mutations is reflected by a selective value loss of MDR strains, Bayesian inferences show that the strains of the Asian lineage (CC1-MDR) have a mathematical growth factor 10 times higher than that of sensitive strains. Another parameter that can be inferred from the coalescent is the lineage expansion dates. Remarkably, even though the CC1-S expansion dates back to approximately 250 years ago, the antibiotic resistant strains started to expand only about 50 years ago. This phenomenon is in line with the democratization of antibiotics in the regions in question.

Admittedly, MIRUs have the double advantage of being a molecular marker for evolutionists and an epidemiologic monitoring tool for doctors. In terms of information completeness, however, they are less holistic compared to data generated by full sequencing of bacterial genomes. Since about 10 years ago, we have been aware that bacteria accumulate mutations relatively constantly. When working with a clonal complex, we often find very recently sequenced strains having phylogenetic trees with branches longer than strains isolated 10 or 20 years ago. By applying a linear regression of the branch lengths and the number of years since isolation, we can obtain a positive correlation, with a slope that is the mutation rate. By applying this strategy to an epidemic that swept through Hamburg from 1997 to 2010, researchers were able to identify a mutation rate at 10^{-7} substitution/ nucleotide/year, which means that a new mutation occurred on a *M. tuberculosis* genome every two years [ROE 13]. During the Hamburg epidemics, genomics revealed two sub-clones with contrasted epidemic success and they are separated by only 20 mutations. It was a rare opportunity to find a genetic determinant that could explain a fitness difference. The focus was set on the *cydD* gene, an *aarD* homolog in *Providencia stuartii* responsible for slowing down the growth rate of this bacterium. The time needed to multiply and to regenerate can be critical for bacterium infectiousness. The fitnesses of two variants (wild and mutant) of this gene were tested in murine models. The experiment unfortunately failed to identify a virulence difference in mice. Nevertheless, the information gleaned from the sequencing of about 100 strains helped us to reconstruct the evolution of the effective population size of the Beijing strains (Figure 8.2). This figure globally shows two main growth increments, with a first surge around the industrial revolution and a second one at the end of the 19th Century. After a demographic stabilization phase, a first inflection is observed upon the arrival of antibiotics in the 1960s, followed by a slight rebound with the emergence of AIDS.

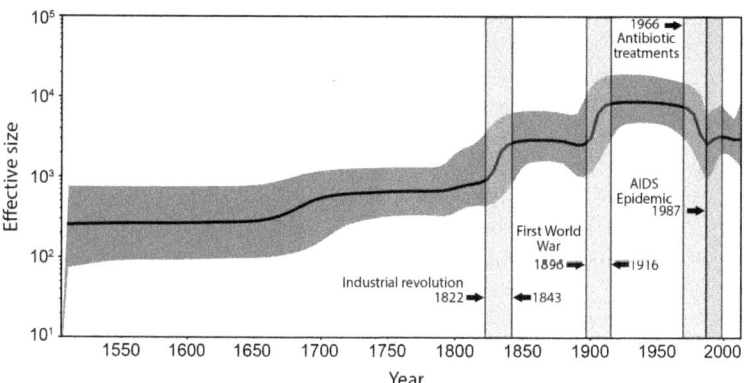

Figure 8.2. *Bayesian analysis of the effective population size fluctuations of the Beijing* M. tuberculosis *lineage. The dark gray areas correspond to the confidence intervals (95%), and the light gray vertical columns represent the main socio-economic events associated with the demographic fluctuations of tuberculosis*

Another objective of this study was to better understand the significant contrasts observed between the different Beijing CCs, for example MDR rates drop from 75% to 5% when considering, respectively, CC2 and CC5. In general, there are two possible strategies to detect genes undergoing diversifying selection. One is to identify recurrent mutations (known as homoplasies) on several independent tree branches; the other is to seek canonical mutations in some emerging lineages. By combining these two approaches, we were not only able to list about 50 new mutations related to antibiotic resistance (*mce* and *VapC*), but also genes involved in the metabolism of lipids or in the exposure of surface proteins. One of the most remarkable examples is the reading frame interruption of the Russian clone's *kdpD* gene, which is involved in the suppression of *M. tuberculosis* multiplication *in vivo*.

8.4. *Staphylococcus aureus* and globalization

Staphylococcus aureus has long been recognized as one of the most common bacteria causing diseases in humans. It is the main cause of skin and soft tissue infections such as abscesses, boils and cellulitis. Although most *S. aureus* infections are not serious, *S. aureus* can indeed cause serious infections, such as septicemia, pneumonia, as well as joint and bone infections.

8.5. Intensive farming and its distorting practices

Both globalization and industrialization facilitate the transmission of human diseases to livestock. Researchers from the Roslin Institute at the University of Edinburgh showed that a *S. aureus* strain of human origin was transmitted to poultry raised in chicken farms [LOW 09]. This work provided the first clear evidence of this type of transmission. The study identified a form of the *S. aureus* bacterium – of which methicillin resistant *S. aureus* (MRSA) is a sub-type – in chicken, and confirmed that they originally came from humans. Genetic analyses and molecular calibration allowed researchers to date this interspecies transfer back to about 40 years ago, which corresponds to the transition from smallholder poultry production to intensive industrial farming. Unlike the ancestral human strain, which is relatively confined geographically, the poultry variant is spread over different continents. Infectious poultry disease in intensive production systems is a major economic burden for the industry, and bacteria jumping from humans to animals could have huge negative impacts. Unfortunately, cases of bacterial pathogens crossing over from humans to animals continue to be reported every year, *Escherichia coli* is yet another case which highlights an increasing threat of human–animal transmission on food safety. Dr Ross Fitzgerald, from the Roslin Institute, stated: "A century ago, chicken were raised for eggs, as meat was considered a by-product. Nowadays, the demand for meat resulted in a poultry industry dominated by a few multinationals, which provide a limited number of reproduction lineages on a world market – thus promoting the propagation of bacteria throughout the world".

This type of interspecies transmission chain is also illustrated by another *S. aureus* strain, called CC398. Similar to the previous scenario, an MRSA, which was originally sensitive to methicillin and which was frequently found in European industrial farms, is also found to have human origins [PRI 12]. Treating pigs with antibiotics was a widespread practice in the 1980s and 1990s. The goal was not to cure them from any chronic infections, but to make them grow as fast as possible, as antibiotics help the absorption of nutrients. Consequently, a one-month-old piglet would weigh 90 times more by the time it was slaughtered at 6 months of age. The evolutionary history of CC398 reveals a complex scenario. Its basal strains are of Asian origin (human and poultry), came back to humans in the form of CCC398 resistant to multiple antibiotics, and leading to numerous nosocomial infections. A biogeographic overview of the transmission illustrated the globalization of this disease. It was brought to Northern America from the East, and then returned to Europe.

From the above examples, we can get a glimpse of one of the many negative influences of industrial farming practices on human health worldwide, as well as the potentially devastating effects of insufficiently controlled international trade.

8.6. Success of an American clone

Over the last two decades, community strains of *S. aureus* (CA-MRSA), associated with methicillin resistance, have significantly increased the global burden of *S. aureus* infections. The pandemic sequence type ST8 called USA300 is a prevailing clone of CA-MRSA in the United States. Lowy and his colleagues succeeded in exploring the short-term evolution and transmission of this germ through full genome sequencing of 387 isolates drawn from an epidemiological network of CA-MRSA infections and colonizations in northern Manhattan [UHL 14]. The clone USA300 diverged from a common ancestor around 1993. Since then, there have been multiple introductions of USA300 in the community. The scientists also reconstructed the dispersion patterns of these strains across the neighborhood. Most of the USA300 isolates had become endemic in households, indicating their essential roles as reservoirs of persistence and transmission. On the basis of a maximum genetic variability threshold per household, the authors further identified several possible transmission networks beyond households. Finally, this study also revealed the evolution of a sub-population resistant to fluoroquinolone in the mid-1990s, which was accompanied by a secondary expansion. This example well illustrated the Red Queen hypothesis. In a more recent study, researchers tried to assess the health risk posed by this American clone in Europe [GLA 15]. Still based on analyses of full genomes, these authors estimated that there was at least a dozen of independent transatlantic import to France alone. There is not an explanation yet, but the American strain fortunately did not flourish in Europe. This study highlighted nonetheless the how human mobility facilitated intercontinental transfers of pathogens.

8.7. Typhoid fever

Typhoid fever is a human systemic infection caused by *Salmonella* Typhi. During the acute phase of the infection, patients can experience symptoms including nausea, abdominal pain, headaches and fever. According to the World Health Organization (2016), typhoid fever annually affects 21 million people and causes 222,000 deaths worldwide. Transmission primarily occurs through the consumption of contaminated food and water, and is therefore mainly limited to poor countries with poor sanitary conditions. Patients respond very differently to the bacterial infection, and asymptomatic carriers can play a significant role in the spreading of the disease.

8.8. Indian diaspora and emergence of a new pathogen

Our knowledge of *S.* Typhi recently made a quantum leap thanks to new sequencing technologies. The large-scale work on almost 2,000 strains of a consortium of mainly British and Australian researchers made it possible to reveal the rise of a multi-resistant clone called H58, which is gradually replacing lineages sensitive to antibiotics on an international scale [WON 15]. By reconstructing the diversification tree of this lineage, it appears that the ancestral strains of the clonal complex were only present in India. They then gradually spread to the bordering countries in Southeast Asia, before spreading to Eastern and Southern Africa. Among the most affected countries are Fiji Islands, Kenya and South Africa. These are countries where the Indian diaspora is the most significant [WIR 15]. Thanks to Bayesian analyses and the coalescent theory, researchers managed to date back the emergence of H58 to the end of the 1980s, and detected a 500-fold increase in the effective population size since that time. It is tempting to link the success of these strains to punctual chromosomal mutations, such as those affecting the genes of the DNA gyrase and topoisomerase IV involved in the reduction of the susceptibility to fluoroquinolone. However, horizontal transfers of antibiotic resistance genes are also involved, highlighting the large adaptive range of microorganisms. The relative youthfulness of this lineage (30 years old at most) and the spatial propagation of the H58 clone emphasize the major role of human immigration. Numerous members of the Indian diaspora have close or distant relatives living in India. Their familial business or leisure visits maintain a network of bidirectional connections, which generates a "signature" topology of the H58 clone phylogeny (Figure 8.3).

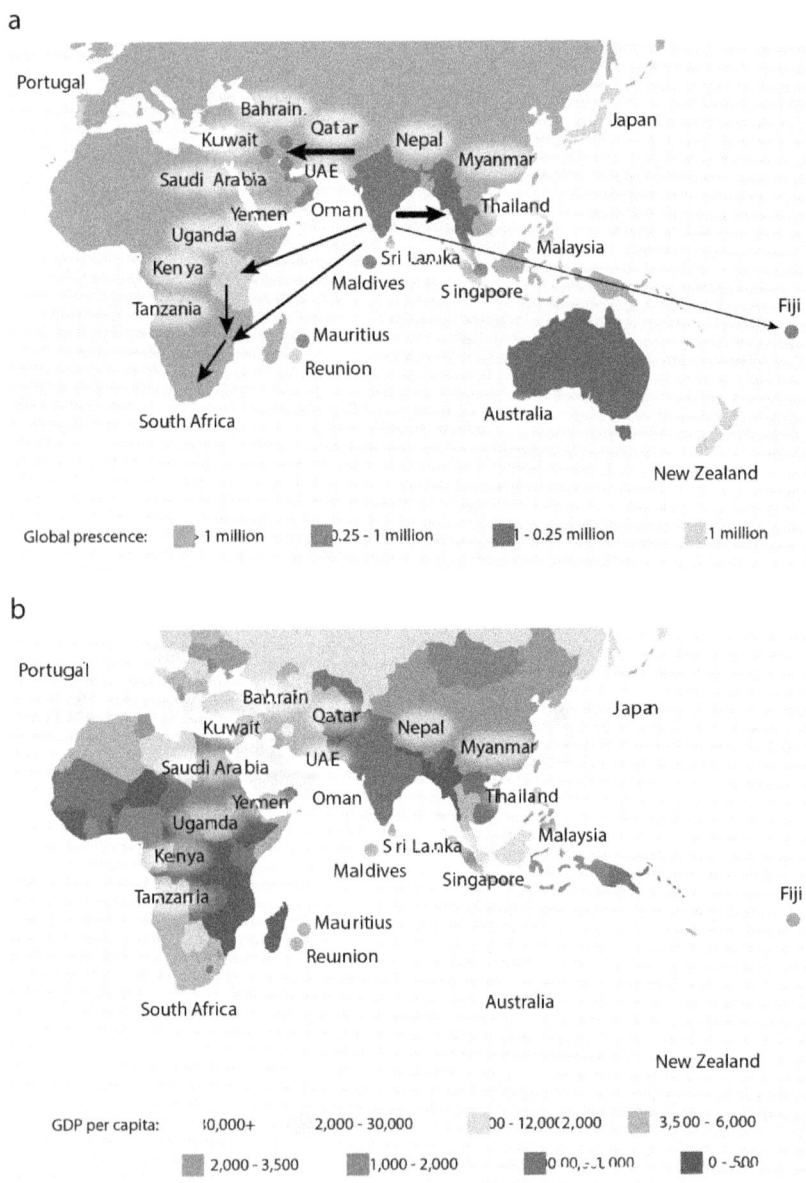

Figure 8.3. *Indian emigration and the standard of living in the countries affected. a) World map of the Indian diaspora. The arrows show the main transmission chains detected by Wong et al. b) GDP in dollars of the countries concerned. Waves of migration as well as sanitary conditions are the main factors, which explain the H58 clone worldwide distribution. For a color version of this figure, see www.iste.co.uk/ grandcolas/biodiversity.zip*

8.9. Prospects

Controlling and preventing epidemics does not only rely on clinical medicine. Statistical tools and modeling can also contribute to the continuing battle by generating indexes of high epidemiologic value. How to define an epidemic strain? An endemic strain? According to which criteria? Without rigorous mathematical proof, any response would risk being conjectural or subjective. The development of the THD (Time-scaled Haplotypic Diversity) index helped to infer the epidemicity of a strain in time and space, and to trace the temporal and spatial dynamic of an epidemic [RAS 17]. By extracting some vectors on acute viral epidemics, this type of tool can produce predictive data on areas of risk that will probabilistically soon be affected. Finally, the expansion of the genetic association at the genome level to traits (THD) other than those linked to antibiotic resistant phenotypes could open up new prospects for understanding what makes a strain successful. Other applications include the possibility to reconstruct the appearance order of mutations in evolutionary lineages. This tool can also help us understand the failure of the DOTS campaign, which was initiated in Uzbekistan in 1998. Due to the presence of numerous mutations allowing resistance to first-line antibiotics prior to the collective effort to treat TB, DOTS only promoted the emergence of MDR strains, which required a second costly campaign in 2003.

Thanks to the relatively universal analytical tools, we are able to understand the evolutionary history of different pathogens and assess their impact on global public health. The three bacterial species mentioned in this short chapter (*M. tuberculosis*, *S. aureus* and *S.* Typhi) are epitomes which illustrate the effects of globalization and themselves consequently. To sum up, understanding a disease always requires assessing the long evolution or coevolution of the pathogen and its host, integrating demographic, socio-cultural and economic parameters, as well as unraveling the contribution of past and present migrations. It captures the complexity of multidisciplinary research.

8.10. Bibliography

[BER 13] BERNARD C., BROSSIER F., SOUGAKOFF W. *et al.*, "A surge of MDR and XDR tuberculosis in France among patients born in the Former Soviet Union", *Euro Surveill*, vol. 18, p. 20555, 2013.

[COM 13] COMAS I., COSCOLLA M., LUO T. *et al.*, "Out-of-Africa migration and Neolithic coexpansion of Mycobacterium tuberculosis with modern humans", *Nat Genet*, vol. 45, pp. 1176–1182, 2013.

[GEN 12] GENGENBACHER M., KAUFMANN S.H., "Mycobacterium tuberculosis: success through dormancy", *FEMS Microbiol Review*, vol. 36, pp. 514–532, 2012.

[GLA 15] GLASER P., MARTINS-SIMOES P., VILLAIN A. *et al.*, "Demography and intercontinental spread of the USA300 community-acquired methicillin-resistant staphylococcus aureus lineage", *MBio*, vol. 7, 2015.

[LOW 09] LOWDER B.V., GUINANE C.M., BEN ZAKOUR N.L. *et al.*, "Recent human-to-poultry host jump, adaptation, and pandemic spread of Staphylococcus aureus", *Proceedings of the National Academy of Sciences of the United States of America*, vol. 106, pp. 19545–19550, 2009.

[MAR 14] MARKS S.M., FLOOD J., SEAWORTH B. *et al.*, "Treatment practices, outcomes, and costs of multidrug-resistant and extensively drug-resistant tuberculosis, United States, 2005–2007", *Emerg Infect Dis*, vol. 20, pp. 812–821, 2014.

[MER 15] MERKER M., BLIN C., MONA S. *et al.*, "Evolutionary history and global spread of the Mycobacterium tuberculosis Beijing lineage", *Nature Genetics*, vol. 47, pp. 242–249, 2015.

[POS 15] POSA A., MAIXNER F., MENDE B.G. *et al.*, "Tuberculosis in late neolithic-early copper age human skeletal remains from hungary", *Tuberculosis (Edinb)*, vol. 1, no. 95, pp. S18–22, 2015.

[PRI 12] PRICE L.B., STEGGER M., HASMAN H. *et al.*, "Staphylococcus aureus CC398: host adaptation and emergence of methicillin resistance in livestock", *MBio*, vol. 3, 2012.

[RAS 17] RASIGADE J.P., BARBIER M. *et al.*, "Strain-specific estimation of epidemic success provides insights into the transmission dynamics of tuberculosis", *Scientific Reports*, vol. 7, p. 45326, 2017.

[ROE 13] ROETZER A., DIEL R., KOHL T.A. *et al.*, "Whole genome sequencing versus traditional genotyping for investigation of a Mycobacterium tuberculosis outbreak: a longitudinal molecular epidemiological study", *PLoS Med*, vol. 10, p. e1001387, 2013.

[UHL 14] UHLEMANN A.C., DORDEL J., KNOX J.R. *et al.*, "Molecular tracing of the emergence, diversification, and transmission of S. aureus sequence type 8 in a New York community", *Proceedings of the National Academy of Sciences of the United States of America*, vol. 111, pp. 6738–6743, 2014.

[WIR 08] WIRTH T., HILDEBRAND F., ALLIX-BÉGUEC C. *et al.*, "Origin, spread and demography of the Mycobacterium tuberculosis complex", *PLoS Pathog*, vol. 4, p. e1000160, 2008.

[WIR 15] WIRTH T., "Massive lineage replacements and cryptic outbreaks of Salmonella Typhi in eastern and southern Africa", *Nat Genet*, vol. 47, pp. 565–567, 2015.

[WON 15] WONG V.C., BAKER S., PICKARD D. *et al.*, "Phylogeographic analysis of the dominant multidrug-resistant H58 clade of Salmonella Typhi identifies inter- and intra-continental transmission events", *Nat Genet*, vol. 47, pp. 632–639, 2015.

[WOR 13] WORLD HEALTH ORGANIZATION, Global Tuberculosis Report, Geneva, 2013.

[WOR 17] WORLD HEALTH ORGANIZATION, Global Tuberculosis Report, Geneva, 2017.

[ZIN 03] ZINK A.R., SOLA C., REISCHL U. *et al.*, "Characterization of Mycobacterium tuberculosis complex DNAs from Egyptian mummies by spoligotyping", *Journal of Clinical Microbiology*, vol. 41, pp. 359–367, 2003.

Why are *Morpho* Blue?

9.1. Introduction

Their large size and the often brilliant blue of their wings put butterflies from the *Morpho* genus among some of the most spectacular insects in South America. Often mentioned in the reports of explorers in the 19th Century (see [FRU12]), they have been prominent in curiosity cabinets and natural history displays and are still today the subject of much commerce. We would logically expect the biology of such sought-after species to be well known and understood, but this is not the case [NEI 08]. Understanding of the origins and evolution of the iridescent blue color in particular is very limited. Why are *Morpho* (often) blue?

The problem of causality in biology can be tackled on three levels: structural, historical and functional [GOU 02]. (1) A *structural* explanation of the phenotype focuses on its physical or biochemical properties, thus answering the question of "how". It also covers the genetic and developmental origin of the observed phenotype by describing in detail its development during the growth of the individual. This proximate explanation of the phenotype has a temporal dimension, the short time of development, and concerns not only the phenotype itself, but also the structures and processes that generate it. This approach can notably detect if

Chapter written by Vincent DEBAT, Serge BERTHIER, Patrick BLANDIN, Nicolas CHAZOT, Marianne ELIAS, Doris GOMEZ and Violaine LLAURENS.

different structures are capable of producing the same phenotype. (2) The evolution of the phenotype (and of its genetic and developmental bases) can also be placed in its phylogenetic (*historical*) context. The study of phenotypic variation on a macro-evolutionary scale makes it possible to test the influence of evolutionary processes and of a wide array of factors on the evolution of phenotypes, while also taking into account the extent to which different species are related, i.e. the overlap of evolutionary history between species. (3) Finally, we can try to detect which evolutionary pressures influence the emergence, continuation or loss of the phenotype. Ecological or behavioral studies (and also biomechanical) allow us to determine in particular the role of selection, natural or sexual, in the evolution of the phenotype (*functional* explanation). In practice, these approaches are often somewhat combined: evo-devo is in most cases developmental genetics compared between related species, thus bringing together the structural and historical approaches; in much the same way, functional morphology combines the study of physical (structural) properties and ecology. Hence, the comparative method makes it possible to test adaptive hypotheses in a phylogenetic framework. However, it is still rare for these three approaches – structural, historical and functional – to be fully integrated [KLI 06]. In this review, we will give an overview of the nature and evolution of *Morpho* coloration across these three levels of analysis. We will attempt to identify the unanswered questions of most interest and propose possible avenues of research to answer them.

Are *Morpho* really blue? In the present case, we are interested in the color of the wings of adult butterflies (we will therefore put aside the egg, caterpillar and pupa stages, none of which – at least in those species of which these stages are known – shows any blue coloration [RAM 14, BEN 16]). Although iridescent blue is rightly associated with *Morpho*, it is not present throughout the genus. In fact, there are several dark species, predominantly brown, sometimes a greyish ocher, or even with a long wing proportion of orange ocher, as in the giant *M. hecuba* (Figure 9.1(b)). There are also three white-winged species, such as *M. polyphemus* (see Figure 9.1(c) and the phylogeny in Figure 9.2). Sometimes, for instance, in *M. telemachus* (Figure 9.1(a)), the blue is not widespread, only very weakly iridescent and much duller. In typically blue species, the coverage of this color varies from one species to another, always fringed by a black part around its edges and often at the base of the wings (Figure 9.1(d), (e), (f), (h), (i)). Even in those species most famous for their iridescent blue, females are often brown or orange (in *M. rhetenor*, for example, Figure 9.1(f) and (g)).

If the question *"why are* Morpho *blue?"* is undeniably interesting, the subsequent question *"Why are certain* Morpho *not blue?"* is just as interesting and makes it possible to investigate the factors involved in the acquisition and/or loss of this coloration. Finally, butterfly wings have two sides, dorsal and ventral, and the latter, excluding very rare cases of developmental aberration, is never blue: all species (see Figure 9.1(h), (i), (j), (k)) have a brownish ventral side (except the three white species), with eyespots varying in size, color and number – these famous marks that look like a vertebrate's eye. The evolutionary origin of this more subtle coloration and of the presence of these eyespots will be briefly discussed.

Figure 9.1. *Diversity of colors in Morphos. a) Male* Morpho telemachus, *b) Male* Morpho hecuba, *c) Male* Morpho polyphemus, *d) Male* Morpho anaxibia, *e) Male* Morpho helenor papirius, *f) Male* Morpho rhetenor, *g) Female* Morpho rhetenor, *h) Male* Morpho menelaus, *i) Male* Morpho helenor, *j) Male* Morpho menelaus *(ventral side), k) Male* Morpho achilles *(ventral side). For a color version of this figure, see www.iste.co.uk/grandcolas/biodiversity.zip*

9.2. Structural explanation: the iridescent blue in *Morpho* is a physical color

Here we aim to identify the structures and physical properties of wings that are responsible for blue iridescence, and the genetic and cellular mechanisms involved in their formation. To approach this problem, we must first define precisely what we mean by "iridescent blue" and identify the visual properties we want to explain. We can thus define several parameters to describe visual appearance; these can be measured thanks to a spectrophotometer and compared between individuals or species (see Box 9.1). (1) *Hue.* Why do the wings of *Morpho* species reflect light at the wavelengths between 450 and 490 nm – corresponding to blue? In general, blue in animals is rarely caused by pigmentation – blue pigments are more complex, unstable and energetically costly than other pigments (e.g. [BUL 04]; see [BAG 07] and [UMB 13] for review). In fact, the origin of the blue in *Morpho* is not pigmentary but structural, i.e. linked to properties of the scales' surface (we will later see how pigmentation still plays a role). (2) *Brightness.* Some *Morpho* are bright while others are darker. (3) *Glossiness.* While some *Morpho* are noticeably glossy (e.g. *M. cypris*), others are more matte (e.g. *M. anaxibia*). (4) *Saturation.* For a similar hue, certain *Morpho* show a more intense, saturated blue. (5) *Iridescence.* If a *Morpho* is manipulated, and the angle from which it is illuminated and/or observed varies, its color changes, ranging from blue to violet, or even to green and orange. Iridescence is defined as the change in color with the angle of illumination or observation (see Box 9.2 for a description of the physical origin of iridescence). The dominant wavelength being that of blue, *Morpho* wings are thus generally seen as iridescent blue.

The initial question of the origin of blue in *Morpho* therefore becomes more specific through the different parameters describing the visual impression generated, such as hue, brightness, glossiness, saturation and iridescence (see Boxes 9.1 and 9.2). We have to explain the physical origin of these different visual properties, and also what makes them vary between species: what structural differences explain these different properties?

Different parameters of color can be illustrated using a reflectance spectrum, which gives the amount of reflected light as a function of the wavelength (Figure 9.2).

Hue – color in the common meaning of the term (blue, green, yellow, red). It often corresponds to the maximum value in the reflectance spectrum of the object.

Brightness – average reflectance (average proportion of reflected light), corresponding to the level of grayness of a color (i.e. gray-spotted space under the curve or line in Figure 9.2).

Saturation (or Chroma) – characterizes the color's purity. It is linked to the spectral length of the object's peaks of reflection or transmission: the broader the peak, the less pure the color, and vice versa (see yellow arrow in Figure 9.2). It ranges from 1 for a monochromatic light (laser) to 0 for a white color.

Iridescence – change of hue as a result of change of angle of observation and/or illumination (i.e. here, shift in reflectance peak quantified by the purple arrow in Figure 9.2). The physical origin of iridescence is shown in Box 9.2.

Glossiness cannot be quantified with one reflectance spectrum alone and requires more complex methods of study. It corresponds to reflection in a particular direction (anisotropy). Matte surfaces, on the contrary, reflect light in all directions (isotropy).

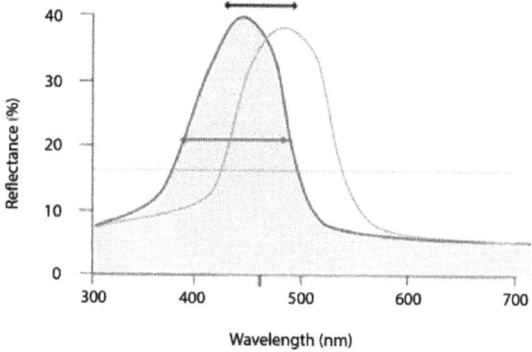

Figure 9.2. *Reflectance spectra as a function of wavelength, illustrating how different parameters of color can be computed: the position in wavelength of the main peak defines hue (here, blue, between 450 and 490 nm); peak width at half height (yellow arrow) defines saturation; the surface under the curve, or the average reflectance (dotted) defines brightness; and a change in curve position (moving from the blue curve to the green curve: purple arrow) as a function of the angle of lighting or observation, defines iridescence. For a color version of this figure, see www.iste.co.uk/grandcolas/biodiversity.zip*

Box 9.1. *Different parameters of color*

Iridescence (also called goniochromism) is a color change of an object as a result of the angle of observation and/or illumination (Figure 9.3).

Figure 9.3. *Iridescence of* Morpho menelaus. *Here, the angle of observation is fixed (perpendicular to the wing) and only the angle of lighting is varied. The values in brackets indicate respectively height and azimuth. For a color version of this figure, see www.iste.co.uk/grandcolas/biodiversity.zip*

Iridescence is caused by two distinct mechanisms that work in tandem in *Morpho*:

1) Thin-film interference – this is the most widespread phenomenon in nature. It results from the reflection of light on two sides of a thin transparent film. When these two waves are in phase, the amplitudes add up. They cancel each other out when they are out of phase. A simple trigonometric calculation tells us this happens when:

$$k\lambda = 2ne\cos\theta_r$$

where k is an integer, λ the wavelength, e the film thickness, n its refraction index and θ_r the angle of refraction (see Figure 9.4). As the cosine is a decreasing function between 0 and $\pi/2$, wavelength λ decreases as the angle of incidence increases and the color seen shifts toward blue ("blue shift"). In *Morpho*, this iridescence by interference takes place mostly on the basal scales, on the level of the corrugated lamellae (Figure 9.6 on right). It is important to note that thin film interferences are a phenomenon of reflection: the angles of illumination and observation must be varied simultaneously and by the same value.

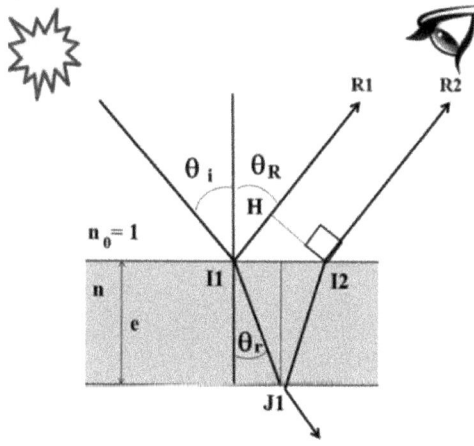

Figure 9.4. *Thin stratum interference. According to Descartes, θ_i and θ_R are equal; to observe iridescence, the angle of observation must be changed as the angle of incidence of the light varies. For a color version of this figure, see www.iste.co.uk/grandcolas/biodiversity.zip*

2) Diffraction by grating – a grating is a grouping of objects (lines, gaps, etc.) regularly arranged. Each of these objects refracts incident light and only certain wavelengths are in phase, and therefore visible, in certain directions. Calculations show us that this occurs when:

$$nk\lambda = \sin i + \sin i'$$

where k is an integer, n the number of objects per meter (the opposite of the step of grating, a), i the light's angle of incidence and i' that of observation (see Figure 9.5). We now have a sine law, a function that increases between 0 and $\pi/2$, that shows us red hues are observed at larger angles than blue ones ("red shift"), contrary to interference. This phenomenon occurs in *Morpho* on the level of the striations that run along the length of the scales (Figure 9.6) and act as a diffracting pattern. We should also note that, in this case, iridescence is only observed by changing the angle of observation.

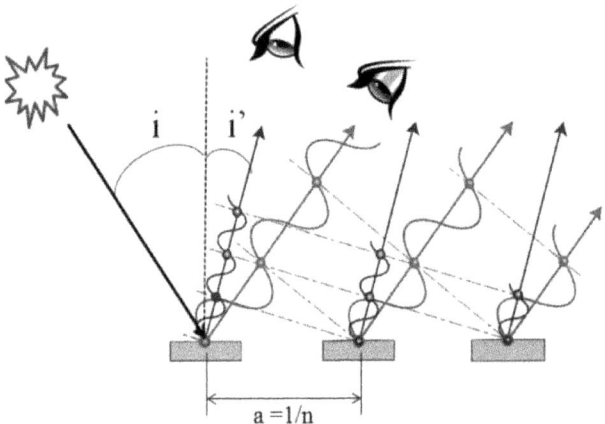

Figure 9.5. *Diffraction by a grating of spacing a. Here we have only shown an order of 1 (k = 1); the same phenomenon occurs at larger angles (k = 2, 3...). For a color version of this figure, see www.iste.co.uk/grandcolas/biodiversity.zip*

The iridescence observed in *Morpho* is therefore complex, since it combines these two antagonistic effects. Each scale acts as a striation pattern whose spacing, in the range of 1 µm, depends on the species and causes a color shift towards red (Figure 9.6, left). But each striation is made of corrugated lamellae roughly 100nm thick that produce the color blue, and also a shift towards blue when the angle of incidence and observation increases.

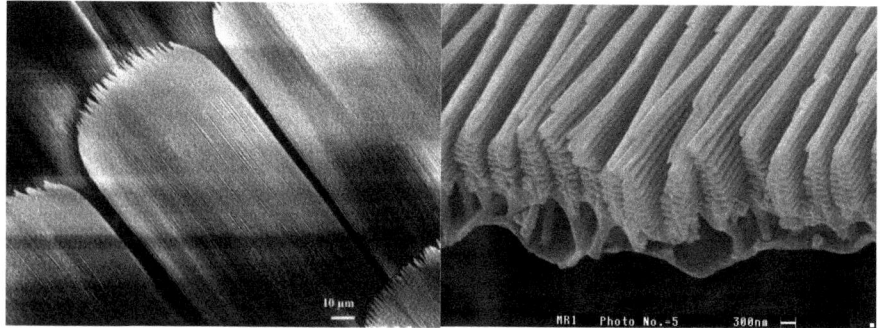

Figure 9.6. *On the left, basal scale of* Morpho menelaus *showing the striation pattern, with a step of approximately 1μm. On the right, corrugated blades (known as "Christmas trees", see also Figure 9.7) producing blue interferences*

Box 9.2. *Iridescence*

9.2.1. *Scale structure*

Butterfly wings, like that of other insects, are made of two cellular layers on top of one another: the dorsal layer and the ventral layer. What is particular to butterflies (and by definition to all Lepidoptera) is that the ventral and dorsal sides present two layers of scales that more or less overlap, called cover scales and basal scales (e.g. [NIJ 91]). The relative size of these scales and the degree of overlap of these two layers are variable between species of *Morpho* and can influence the visual properties of their wings [BER 10]. Unlike other iridescent butterflies, the structure of basal scales, and not that of cover scales, produces most of the photonic phenomena in Morpho species (e.g. [ING 08]): thus the light passes through a layer of transparent scales before reaching the iridescent scales (and passes through it again after reflection). The understanding of the optic properties of *Morpho*'s wings has been made possible by the technical progress of microscopy, which has allowed the identification of the micro- and nano-structures responsible for them (e.g. [GHI 72]; see [YOS 04, ING 08, BER 10] for historical descriptions).

Scales are marked longitudinally, from the root to the apex, by microscopic parallel striations (at least on the main part of the scale; Figures 9.6 and 9.7(a–b)). These striations are made of corrugated lamellae of chitin, of varying quantity depending on species (but very constant within each species), slightly tilted from scale surface. Striation combined with lamellae three-dimensional organization (Figure 9.7) is responsible for scale photonic properties, notably iridescence [ING 08, BER 10] (see Box 9.2). In particular, lamellae corrugation amplifies the amount of light reflected, which becomes stronger (the color becomes more saturated) as lamellae number increases [BER 10, GIR 16]. Thus, intensely colored species such as *M. rhetenor* and *M. cypris* show striations made of ten to twelve corrugated lamellae, while striations in *M. helenor*, which displays a less striking blue, only contain three or four corrugated lamellae [GIR 16]. This corrugation is also the cause of the blue coloration, through an effect of interference (see Box 9.2 and Figure 9.7(d)). In fact, each lamella partly reflects the light that reaches its surface: each successive ray of reflected light interferes, sometimes positively (which increases the amplitude of certain wavelengths, here those in the blue range) and sometimes negatively (which reduces the amplitude of other wavelengths). Distance between lamellae (in "Christmas tree" organization, see Figures 9.6 and 9.7(c)) plays a large part in determining the wavelength reflected (and therefore the hue).

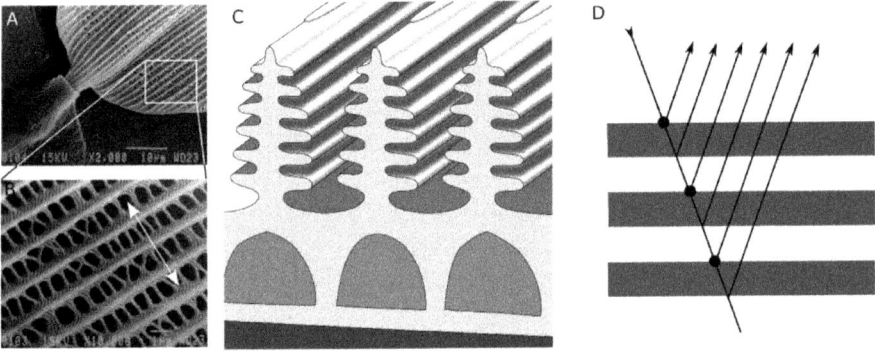

Figure 9.7. *Micro- and nano-structure of scales responsible for blue iridescence. a) View in electronic microscopy of scale base showing the pedicel. The longitudinal striations covering the scale are clearly visible. b) Detailed view of striations under strongest magnification (electronic microscope). The double-ended arrow indicates the plane of the cross-section corresponding to (c). c) Schematic representation of perpendicular cross-section of a* Morpho *scale. The overlap of chitin lamellae results in this "Christmas tree" organization (see also Figure 9.6). d) Illustration of the phenomenon of reflection/positive interference, linked to the superposition of lamellae separated by air. (a) and (b) from [BER 10]*

As well as their peculiar physical structure, the basal scales contain pigments. Their dark pigmentation (melanin) reduces unwanted light reflection and thus increases color saturation [ING 08; BER 10, GIR 16].

This pigmentation is absent in the two most basal species of the *Morpho* phylogeny, *M. marcus* and *M. eugenia*, in the three white species (*M. polyphemus*, *M. epistrophus* and *M. iphitus*) and in species from the *sulkowskyi* clade, apart from *M. absoloni* (i.e. *M. aega*, *M. portis*, *M. aurora*, *M. rhodopteron*, *M. zephyritis*, *M. lympharis* and *M. sulkowskyi*) (see Figure 9.7; [BER 10]). This absence of melanin contributes to the generally less intensely blue appearance of these species.

Glossiness diversity of *Morpho* species (comparing, for example, the very glossy *M. rhetenor* or *M. sulkowskyi* to the more matte *M. helenor* or *M. menelaus*) is influenced by the cover scales [VUK 99, YOS 04]. In the glossiest species, cover scales are either reduced or provide little coverage, while they are more developed in other species, where they may act as an "optic filter", reducing wing glossiness while producing more diffuse reflected light [YOS 04] (see Box 9.1). Scale shape can also play a role: in *M. anaxibia*, a species of a deep but quite matte blue, scales are convex, which reduces glossiness [BER 10].

Two species, *M. eugenia* and *M. marcus* – the most basal species of the group – show optical properties that stem from fundamentally different causes than other species of *Morpho*. In these two species, scale striations are made of only one chitin lamella: iridescence is therefore not produced by the corrugation of lamellae but by that of the scales themselves [BER 10]. These two different organisations have very little influence on hue but directly affect the spatial distribution of reflected waves. Distance between scales is far greater than the length of the wave packet, or coherence length, which is of circa 1μm for solar light. This prevents any coherent effect, in particular diffraction by the striation pattern. We can therefore suppose that iridescence appeared in *Morpho* twice independently early during their diversification.

There is another way of producing blue colors in nature: fluorescence (see [LAG 15] for review). By a nonlinear process, high-energy radiations (generally UV) are absorbed by fluorescent molecules, and re-emitted at a lower energy, usually as blue or green. *M. sulkowskyi* deserves particular attention in this regard (e.g. [KUM 94]). It is the only species of the *Morpho* genus to show noticeable fluorescence. In this species, the basal scales lack the melanins that absorb ultraviolet rays and limit fluorescence. Instead, they contain purines that fluoresce. The emission spectrum shows a blue peak at 480 nm. The output, while remaining relatively weak, adds a noticeable base of blue to this otherwise very pale species [VAN 11a, VAN 11b].

A final optic property of *Morpho* wings is polarization. In a wave model, light is an electromagnetic wave where electric and magnetic fields are crosswise, i.e. perpendicular to the direction of propagation. Natural light is unpolarized. Fields oscillate in all directions perpendicular to the light ray. Various devices called polarizers select one direction of oscillation. Any wave that has undergone an oblique reflection, for example on foliage or on water surface, is partially polarized. This is also the case for light reflected by the scales of certain *Morpho*, as a result of raised striations of lamellae [VUK 99, BER 10]. Imperceptible to the human eye, this effect is interestingly visible to various insects (e.g. [KEL 99]), such as butterflies. Kelber [KEL 01] showed that butterflies use polarization, combined with color information, to choose their site of oviposition (see also [DOU 07] for a discussion on the biological role of polarization). The ecological importance of polarization for *Morpho* is unknown.

9.2.2. *Scale development*

If the structural origin of *Morpho* coloration has been relatively well identified, the genetic and development origin of the structures involved are less well known. Colors of butterflies are, however, the subject of numerous research programs, notably in evo-devo (e.g. [ALL 08, BEL 02, HEL 12, JOR 11, LEP 14, NAD 16]). However, most genetic and developmental studies are interested in pigment synthesis as well as the positioning of color patterns and contain little or no information on scale structure and development (e.g. [BEL 02, NAD 16]). Yet, it has been suggested that these two aspects, pigmentation and structure, are linked (e.g. [GIL 88]). There is therefore little work available on the development of butterfly scales [OVE 66, GHI 02, GHI 76, GAL 98, CHO 12, DIN 14].

It has been shown that scales are homologous with the sensory bristles of other insects (notably of Drosophila; [GAL 98]), the development of which is well studied (e.g. [SIM 90, SKE 91]). Galant *et al.* [GAL 98] showed that, in the *Precis coenia* (Nymphalidae) butterfly, the scales are formed at the start of pupation by two waves of cellular division: the first, followed by massive apoptosis, organizes the cells in successive rows along the proximo-distal axis of the wing; the surviving cells undergo a second wave of division which in the case where each dividing cell gives rise to one cell that produces the scale and one cell producing the pedicel of the scale ("*socket cell*", Figure 9.7(a)). The similarity with sensorial bristle development can be seen at a morphological level but is even more striking at the genetic level (notably by the expression pattern of the *achaete scute* homologous gene, a gene playing a central role in the differentiation of bristles in Drosophila (see [GAR 09] for review). These similarities support the hypothesis that scales are modified sensorial bristles, co-opted by Lepidoptera in the development of colored scales [GAL 98]. The scale itself is formed from skeletal cellular material; microtubules undergo special growth at the end of wing development and form fiber bundles that are particularly important to the formation of the striations on scale surface [OVE 66]. In a recent study, Dinwiddie *et al.* [DIN 14] showed that actin filaments play an important role in scales morphogenesis, and in particular in the development of dorsal striations that cause most of the optic phenomena discussed. Finally, ploidy might also play a role in scales development: cells that produce scales are polyploid, and the ploidy level is, at least in *Manduca sexta* (Sphingidae), correlated with scale size [CHO 12].

Certain aspects of the three-dimensional structure of *Morpho* scales described here, in particular the corrugated lamellae in "Christmas tree" organization, are also present in other butterfly species, but on the cover scales and not the basal scales (see [VUK 00]). This is the case for numerous Pieridae species (e.g. the genera *Eurema, Colias* or *Gonepteryx*; [GHI 76, WIL 11]). These are not blue, but they nonetheless show a significant component of physical color in the ultraviolet range (e.g. [GHI 76, WIL 11]). Similarly, the three-dimensional structure of scale striations in *Trogonoptera brookiana* (Papilionidae) is very similar to that of *Morpho* [WIL 16].

Iridescence may be adaptive in different contexts, but can also evolve in a neutral way from non-iridescent scales of simple structure. In an artificial selection experiment on *Bicyclus anynana* (Nymphalidae), Wasik *et al.* [WAS 14] obtained individuals with partially iridescent wings in the violet range after only six generations of artificial selection from brown-winged ancestors. This study highlights the presence of genetic variation for this trait in this species. Several species in the genus *Bicyclus* show violet iridescent scales, the evolution of which seems to have occurred independently and by different means (iridescence of basal scales or cover scales according to the species). Wasik *et al.* [WAS 14] therefore suggested that the adaptive evolution of structural colors may be easier than that of pigmentary colors: unlike the latter, which often requires diet changes to obtain new pigment(s), changes involved in the evolution of structural colors are of a quantitative nature, occurring by modulation of the quantity of chitin secreted by wing cells. Finally, Ghiradella and Radigan [GIR 76] suggested that self-organizing processes in intra-cellular structures could play an important role in the development of butterfly scales, which could explain why few genetic changes are necessary to modify their structure, and therefore their color.

9.3. Historical explanation: evolutionary origin of blue color in *Morpho*

Morpho are part of the Satyrinae sub-family, where they form, along with *Antirrhea* and *Caerois*, the *Morphini* tribe [DEV 85, PEN 06], the sister tribe of the *Brassolini* (including the renowned owl-butterflies from the *Caligo* genus) [FRE 04, WAH 09]. The *Antirrhea* and *Caerois* do not have the spectacular iridescence of some *Morpho*, but they do quite often show zones of violet–blue iridescence (personal observation; see [DAB 84]). These butterflies fly in the understory, mostly in shadow (never in environments fully exposed to sunlight), and most often at the ground level [DEV 85]. Their ecology seems quite different from that of *Morpho*, which comprise high-flying species, flying in or above the canopy, and species flying mainly in the understory [MIC 11, DEV 10, CHA 16]). The *Morpho* genus comprises 30 species, for which numerous sub-species have been described (e.g. [LAM 04, BLA 07, BLA 12]). Their phylogeny has been the subject of much work in recent years [PEN 02, CAS 10, CAS 12, PEN 12, BLA 13, CHA 16] and is now well established (Figure 9.8 modified according to [CHA 16]).

9.3.1. *Color variation in the genus* Morpho

The most basal lineage of the genus comprises two species, *Morpho marcus* and *M. eugenia*, which both show an iridescent blue color in males, while the females are brown-black with a yellow band across the two wings (Figure 9.7). These two species are described as flying mainly in the understory [PEN 02, CHA 16]. The rest of the genus is then divided into two clades: a clade with canopy-flying large species with lengthened triangular forewings, a trait suspected to be an adaptation to gliding in open spaces [DEV 10, CHA 16], and the second clade being generally associated with flight in the understory. The "canopy clade" is split in two: first, a group of large to very large butterflies whose flight pattern is typically gliding, and in which the iridescence characteristic of the genus is either reduced or absent (the "*telemachus*" group; see Figure 9.1, *M. telemachus* and *M. hecuba*; see also Figure 9.7); second, a group of three species, *M. anaxibia, M. cypris* and *M. rhetenor*, all of which are blue, and the latter two highly iridescent. These three species are sexually dimorphic in color: in *M. anaxibia* females, the blue covers less surface than in males; in *M. rhetenor* females, the dorsal side is ocher-orange with brown-black patterning (Figure 9.1(f) and (g)); the same is true of *M. cypris*, a species in which there are, however, females with the brilliant blue characteristic of males on part of their wings.

In the "understory clade", color diversity is also large. A white species, *M. polyphemus*, forms the base of the clade, which is divided into two sub-clades. The first sub-clade itself contains two groups, one made of three relatively large blue species, *M. amathonte, M. menelaus* and *M. godartii*, all three of which are fairly dimorphic, the females showing brown-black edges with clear marks (and therefore a smaller blue area than in males). The second group contains eight small species, often very glossy, but whose hue is a pale blue, and even slips into white in certain species, such as *M. sulkowskyi*. Two species in this group show particularly remarkable sexual dimorphism: *M. aega*, the females of which are polymorphic, one forming an ocher-orange color with brown-black patterning, another with a large covering of iridescent blue and a third intermediary form; and *M. zephyritis*, the females of which are ivory with very weak iridescence, while the males are particularly shiny blue. *Morpho absoloni*, the males of which are also a particularly glossy blue, are unique, differing from other species of the clade as the females have large brown-black margins, while the proximal half of the wings are blue.

The second understory sub-clade contains the type species of the genus *Morpho achilles*, and is divided into two groups, one comprising *M. achilles, M. helenor* and *M. granadensis*, three species characterized by two black parts, proximal and distal, whose width varies geographically. Hence, the blue patch between the black parts can be reduced to a very narrow band (see Figure 9.1(i)), or it can almost completely cover the wings. The second group contains three species, of which two are white (*M. iphitus* and *M. epistrophus*) and resemble the basal species *M. polyphemus* very closely. The third species in this group is *M. deidamia*, whose dorsal side looks quite similar to that of *M. helenor, M. achilles* and *M. granadensis*, with geographical variations, while the ventral side shows a more complex color pattern. This second understory sub-clade is characterized by a very limited color dimorphism, and a fairly consistent yet definitively less glossy and less iridescent blue than in more striking species such as *M. rhetenor* [BER 10].

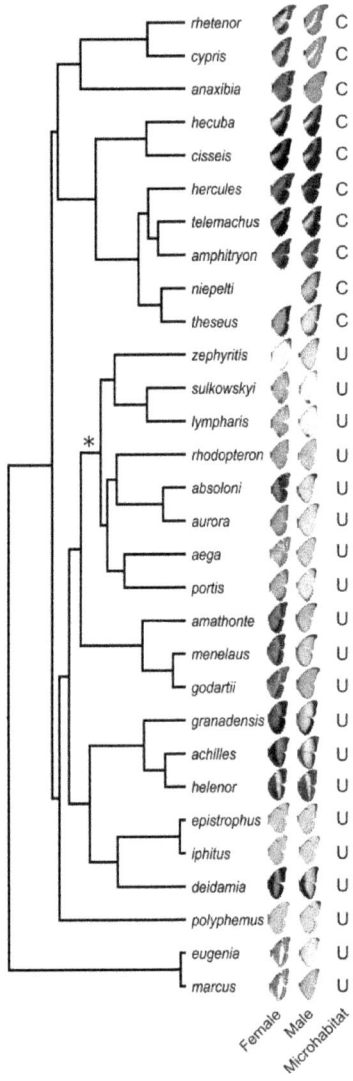

Figure 9.8. *Phylogeny of the genus* Morpho, *showing the right wings of males and females. U = Understory and C = Canopy. Figure modified according to [CHA 16]. For a color version of this figure, see www.iste.co.uk/grandcolas/biodiversity.zip*

This description of color variation on the scale of the genus is inevitably schematic and does not take into account the numerous variations, notably on an intra-specific scale (i.e. the existence of various sub-species and geographical variants). It underlines the inter- and intra-specific diversity (sexual dimorphism) of coloration in *Morpho*, in the clades that live in the understory as well as in the canopy, suggesting a complex evolutionary history of coloration in this genus.

What we can take, however, from this brief description, as well as from looking at the phylogeny (Figure 9.8), is an idea of the importance of phylogeny in determining coloration. In other words, phylogenetically close species seem, in general, to be more similar in terms of coloration (this is called phylogenetic signal). This amounts to the null hypothesis of neutral morphological divergence along the branches of the phylogeny (the Brownian model; e.g. [GAR 94]). In answer to the initial question, *"why are* Morpho *blue?"*, this model would answer *"because they inherit their blue color, or the absence of it, from their ancestors"*, without suggesting any particular adaptive mechanism. But does phylogenetic history suffice to explain the precise distribution of color in this genus? In particular, the evolution toward a very clear, even white, color is intriguing. This has occurred several times, likely independently: other than the white species, species belonging to different clades (*M. theseus, M. godartii, M. aurora, M. helenor*) contain very pale or white sub-species. These colorations may have been favored by ecological factors. Similarly, some species show a marked blue coloration, in contrast to their closest relatives (e.g. *M. absoloni* vs. *M. aurora* or *M. zephyritis* vs. *M. sulkowskyi*), which suggests natural selection may play a role in the evolution of this color. These hypotheses must be explicitly tested, either experimentally or in a comparative framework. Finally, the evolutionary origin of iridescence itself is yet to be explored. The study of wing structures responsible for iridescence and the study of their diversity among *Morpho* species would allow us to identify phenomena of convergence. This could also shed light on the ecological implications of this trait and on selection pressures affecting its evolution (see [BER 10, GIR 16] for the comparative analysis of some of the species of the genus).

9.4. Functional explanation: the role of selection in the evolution of *Morpho* color

The above discussion demonstrates that answering the question "*why are* Morpho *blue?*" also implicates the identification of the selective processes involved in the evolution of this color and its diversification.

Color diversity in animals is often discussed in terms of signals received by predators and/or sexual partners or competitors (e.g. [GOM 07]). Butterfly coloration is therefore generally considered first and foremost as a visual signal, both on an inter- and intra-specific level (see, for example, [END 78, SIL 84, NIJ 91] for general references). Other selective pressures act on coloration: butterfly wing scales are also involved in thermoregulation and hydrophobia.

9.4.1. *Thermoregulation*

In butterflies, the muscles involved in flight must reach a fairly high temperature to work properly, making flight possible (e.g. [KIN 85]). Thermoregulation is therefore particularly important for these insects. Coloration, because of its light-absorbing qualities, plays an important role in thermoregulation (e.g. [WAS 75, KIN 85]). However, this role is generally limited to the base of the wings, suggesting that apical patterns have little to do with thermoregulation ([WAS 75]; see [KIN 85, KIN 87, KIN 95] for a discussion on Pieridae). All variations in color patterning can therefore probably not be explained by their effects on body temperature. The possibility that iridescence may have a particular effect on thermoregulation is quite controversial [DOU 09], certain authors suggesting that high reflectivity tends to reduce the capacity for thermic absorption (e.g. [KOO 00]), others claiming on the other hand that scale structure and organization allow heat to be more effectively directed toward the veins and the hemolymph (e.g. [TAD 98]; see [WAS 75] for further discussion on the role of the hemolymph in thermoregulation). Data gathered about *Morpho* wings suggest that their high reflectivity has little effect on their absorption capacity [BER 10]. If it seems unlikely that blue iridescence in *Morpho* has evolved primarily in response to selective pressures linked to thermoregulation, it is nonetheless within the realms of possibility that the thermal effects of iridescence have played a role in its evolution. This hypothesis should be tested with greater accuracy by comparing the distribution of thermal absorption throughout the phylogeny in tandem with coloration distribution (see [BER 10]). Moreover, literature on thermoregulation is for the most part focused on species in temperate climates [KIN 85], and tropical butterflies are therefore not at all understood in the context of thermoregulation.

9.4.2. *Hydrophobia*

When it rains, butterflies hide under cover, generally on a tree-trunk or under a leaf. In humid tropical climates, it is definitely more difficult to avoid getting wet than in temperate climates. The scales covering butterfly wings in much the same way as roof tiles generally create a very hydrophobic surface (e.g. [WAG 96]). Zheng *et al.* showed that in *Morpho* the existence of raised, overlapping chitin lamellae (Figure 9.3) increases hydrophobia to an extreme degree (see also [BER 10]). In addition, this super-hydrophobia is combined with directional adhesion, where water droplets are prevented from rolling toward the body and instead roll off wing outer edges [ZHE 07]. It is thus conceivable that the evolution of *Morpho* scales nanostructure has been influenced not only by their photonic properties, but also by natural selection favoring more hydrophobic morphologies. Nonetheless, *Morpho* do not usually fly in wet weather. They rather stay under vegetation with their wings closed, where the non-iridescent ventral side of their wings is exposed to the rain.

9.4.3. *Signaling to predators: a confusing effect?*

When it comes to coloration as a visual signal, it is difficult to identify the relative importance of the various visual parameters such as hue, iridescence, brightness, glossiness and saturation, all of which are in reality combined. Analysis of their covariation – or, on the other hand, their potential independence – would be very informative. We will consider these parameters in combination in the following discussion.

Morpho are relatively fast-flying butterflies often considered difficult to capture – at least by human predators ([YOU 71, PIN 96, PIN 16]; personal observation). Their coloration may play a role in making their capture even more difficult, as a result of the marked but intermittent signaling emitted by their wings, irregular and ever-changing in nature as they beat rapidly in flight, hiding and exposing successively the glossy dorsal side and the darker ventral side. The irregular flashes of changing colors (i.e. a dark–light contrast, as well as variation in glossiness and iridescence), combined with a complex flight pattern and escape manoeuvres, make it difficult to locate the butterfly and to predict its flight trajectory. The intense, iridescent and glossy blue may therefore produce sensorial confusion in the predator (see [STE 07] for a general discussion on the sensorial effects of coloration on predators). The hypothesis that iridescence may have an anti-predator role ([ROB 96, HIN 73]; see for review [MEA 09]; see [CLE 66] for particular mention of butterflies) has received little to no experimental testing (see, however, [HIK 15]), and remains completely unexplored in *Morpho* (see [NEI 08] for discussion). Color evolution in *Morpho* would therefore seem to hinge both on light environment and on predator cognitive capacities. In addition, gliding, which seems favored in the canopy [DEV 10, CHA 16], would therefore limit the frequency of "anti-predator flashes", which could explain why iridescent blue is less important in certain canopy species. In any case, these considerations highlight the strong link between the flight evolution and evolution of color patterning.

Besides camouflage, which is a form of signaling limiting detection by predators, the often striking colors of butterflies may indicate to potential predators that an attack would be costly, for example, in the case of toxic butterflies (e.g. [END 88, SHE 08]). This is called an aposematic signal, which confers to those individuals carrying it protection against predators who learn, at their own expense, to avoid them. Avoidance by predators favors aposematism and can lead to convergence on the same aposematic signaling in different toxic species that are exposed to the same predators (Müllerian mimicry). Indeed, toxic species mimicking a signal already known by predators benefit from the protection associated with this signal and reduce even further the individual risk of predation which is spread over a larger number of individuals. Furthermore, non-toxic species can also evolve such aposematic signals, thus benefiting from the aposematic protection associated without entailing the metabolic costs associated with the production of chemical defenses (Batesian mimicry: see, for example, [MAL 99] for review).

The toxicity of butterflies is not the only factor incurring a potential cost for predators. The attack in itself is obviously costly in terms of energy, especially when the butterfly is difficult to capture – for instance if it flies fast or erratically. The energetic cost will be relatively higher for prey with lower calorie content. From the butterfly's point of view, toxicity and difficulty of capture are two traits that limit the risk of being caught by predators.

Box 9.3. *Aposematism*

9.4.4. *Signaling to predators: an aposematic blue?*

Beyond its direct contribution to making them difficult to locate, the blue seen in *Morpho* could equally act as a signal of fast and/or erratic flight to predators, in particular birds. In much the same way as the bright colors of toxic butterflies (see Box 9.3), this color may inform predators about the cost of an attack, but this time in relation to the difficulty of capture rather than toxicity [YOU 71, PIN 96, SRY 99, PIN 16]. This hypothesis has been supported by experimental data: Pinheiro [PIN 96] exposed butterflies to a predator – a Jacamar, an insectivorous bird specialized in catching butterflies – and showed that *Morpho* are very rarely attacked, and in the event of an attack, they are very rarely caught. This conclusion is nonetheless contradicted by field observations concerning *M. rhetenor* ([NEI 08], p. 218) and *M. menelaus* ([GAY 16], pp. 16–17). Data in this area are generally quite rare and the correlation between the blue signal intensity and protection from attack has never been formally tested.

It has also been suggested that this escape aposematism could, like toxicity aposematism, trigger mimetic evolution (referred to as escape or evasive mimicry; [VAN 59]). In other words, if the visual signal associated with the difficulty of capture provides a selective advantage by discouraging attack, convergence toward the same signal could occur among fast flying species (Müllerian escape mimicry). On the contrary, species less difficult to capture could benefit from being similar in appearance to faster species (Batesian escape mimicry). Mallet and Singer [MAL 87] discussed this hypothesis and suggested it could be valid, particularly when the energy benefits of capture are low, such as with big butterflies with a *"thick, indigestible cuticle and solid wings"*, as well as with very small butterflies (low in nutrition), in particular Lycaenidae. This hypothesis has been criticized, mainly because of the lack of robust empirical data, and also on theoretical grounds: Brower [BRO 95] suggested that, unlike toxicity (unpleasant signal), the stimulus that would associate color with failure to catch the prey would not be strong enough to warrant long-term memorizing by the predator, making the evolution of escape mimicry, Batesian or Müllerian, unlikely. Ruxton *et al.* [RUX 05] modeled the evolution of Batesian and Müllerian escape mimicry. Their findings suggest such an evolution is possible in both cases, as long as an attack is costly for the predator, there is abundant alternative prey, and escape is costly for the butterfly.

These hypotheses of *Morpho* color as signaling difficulty of capture, and the subsequent hypotheses of the possible evolution of escape mimicry, have however never been empirically tested. It is nonetheless interesting to note that in *Morpho*, even of the largest species, the body is relatively small and likely poorly profitable as prey.

9.4.5. *Sexual selection*

The coloration of *Morpho* is widely considered to be a form of signaling in intra-specific communication (e.g. [VUK 99]), either toward individuals of the opposite sex in the context of mate choice or toward individuals of the same sex in the context of intra-sexual competition (e.g. [SIL 84]). This hypothesis is not specific to *Morpho*: the coloration of butterflies has generally been interpreted in this context (e.g. [END 78, KEM 11]), and notably by Darwin and Wallace ([KOT 80]; see [SIL 84] for further discussion). This is the case above all in species where there is a strong sexual dimorphism in color, the males being generally more colored, which suggests either sexual selection or ecological differences between sexes associated with different natural selection pressures [ALL 11]. Iridescence in particular has been the subject of much attention, because it allows for directional signaling. The polarization of reflected light has also been suggested as a way of producing easily detectable signals in situations where there is little light, for example, in the understory (e.g. [DOU 07]). It has been shown, in the neo-tropical butterfly *Heliconius cydno* (Nymphalidae), that the polarized iridescent blue of the wings is involved in signaling in sexual encounters [SWE 03]. What about *Morpho*?

Gomez and Théry [GOM 07] showed that light, yellow, white or blue signals create a strong visual contrast (for birds but also more generally for any tetrachromatic system), making it easier to communicate in the understory. They also suggest that a dark saturated blue is more visible in the canopy. These results are in accordance with the distribution of brightness of blue between canopy and understory species: indeed, the three blue canopy species (*M. anaxibia*, *M. rhetenor* and *M. cypris*) show an intense blue that is generally darker than that of understory species (see Figure 9.7), reinforcing the idea of coloration as a visual signal. In blue species, there is also a more or less marked sexual dimorphism in color, where the males always have a more intense blue that covers more of the wing relative to females ([CHA 16]; see Figure 9.7; note that these differences are not quantified, and there is no quantitative data allowing us to compare iridescence and glossiness between sexes). This dimorphism is the strongest in *M. rhetenor*, *M. cypris* and *M. aega*. In the latter two, the females are polymorphic, the predominant morph being an ocher-orange color. It is of interest that color dimorphism is particularly low in non-blue species, which suggests that the blue color is either targeted by sexual selection or by a particularly contrasted natural selection between sexes. Certain species of *Morpho* are often said to be territorial, patrolling in regular fashion a specific area from which they "chase" other males fairly aggressively (e.g. *M. amathonte* [YOU 73]). This territoriality should, however, be considered with caution, as it is generally impossible to follow butterflies in the forest, which limits our understanding of their actual movements and forays into open spaces (rivers, paths and forest borders). In some species (e.g. *M. helenor*), on the contrary, it is not uncommon to observe several males feeding on the same fruit. Nonetheless, in all blue species, the males are attracted by the lure of metallic blue, which suggests either a territorial behavior, or more simply, an inability to discriminate from a distance the lure from a female (in species showing little dimorphism) or from a potential rival. Sex ratio is unknown for most species, but females are much more rarely observed than males, which may indicate a sex ratio that is biased in favor of males (difficult to explain from an evolutionary standpoint) or simply a behavioral difference, males being more visible due to their patrols in open parts of the forest, while females might remain in the close vicinity of host plants to lay eggs (but again, our understanding of behavior is very limited).

The hypothesis of sexual selection by females to explain the evolution of iridescence and sexual dimorphism of coloration has been put forward for some species with iridescent males (e.g. [CON 07]), and validated in experiments (notably for *Hypolimnas bolina* (Nymphalidae), *Eurema hecabe* (Pieridae) [KEM 07a, KEM 07b, KEM14] and *Bicyclus anynana* (Nymphalidae) [ROB 05]). To be demonstrated in *Morpho*, this hypothesis must establish that females choose males differently according to their color, which must therefore be more than a mere signal of species or sex recognition [KEM 11]. Sexually selected traits are generally costly and condition dependent (i.e. dependent on individual health condition, which is in turn linked with genetic quality (see [HIL 11] for discussion); e.g. [ALL 11]). We currently have no data on the condition dependence of coloration in *Morpho*. Kinoshita *et al.* [KIN 02] showed that the irregularity of lamella corrugation formations comprising the striations limits iridescence. High regularity in this area could therefore be important for generating highly iridescent coloration patterns. Indeed, it is imaginable that maintaining such a high regularity is energetically costly during development (see [DOU 09] for review). Variations in the quality of iridescence between males of the same species could thus be linked to variations in their general health (their condition). In turn, iridescence could then be used by females as a mate choice criterion, and their preference could hence evolve through the selective advantage associated with paternal genetic quality transmitted to their offspring.

9.4.6. *Different natural selection between sexes?*

It is also possible that color dimorphism between sexes is linked to different natural selective pressures. This hypothesis is fairly consistent with Wallace's views on sexual dimorphism [KOT 80], which he thought resulted from the appearance of a trait in both sexes followed by its elimination by selection in the most cryptic sex (here, females). We have recently shown [CHA 16] that sexual dimorphism of wing shape was associated with color dimorphism. This association could come from different selective pressures between sexes, females, due to being more cryptic and generally heavier, adopting a different flight pattern from males. In particular, they might fly less than males, leading to reduced predation pressure from birds. This hypothesis goes against Darwin's, focused on sexual selection [KOT 80], here blue being sexually selected by females as an honest signal (i.e. a feature costly to maintain and signaling the good genetic quality of its carrier). However, these two points of view are not exclusive (see [ALL 11] for detailed discussion). In *Morpho,* coloration, if indicative of flight performance and capacity to escape predators, could be directly used as a mate choice criterion by females. On the other hand, coloration could offer no advantage against predators, but on the contrary a handicap, providing females with an indirect criterion of quality.

9.4.7. *And the lack of blue?*

Understanding the evolution of glossy iridescent blue can also be improved by identifying selective pressures that allow for the maintenance of different colorations. First, the ventral face of *Morpho* is not blue but brown-beige with eyespots, the number and size of which vary between species (Figure 9.1). This dorso-ventral contrast strongly suggests that the two faces are subject to different selective pressures – indeed, opposite pressures. If the evolution of high conspicuousness may be favored on the dorsal face by predator behavior, selective pressures on the ventral face seem to promote camouflage. When butterflies are at rest, shaded by foliage with folded wings, the coloration of their ventral side makes them very difficult to spot. Eyespots, more or less visible and marked depending on the species, could divert predator attacks away from the most vital parts (see [STE 05] and [MON 15] on the role of eyespots in other species). It has also been suggested that eyespots may play a role in mate choice (e.g. [STE 05]). These hypotheses have never been tested in *Morpho*.

So what about non-blue species? In the case of the *"telemachus"* group, we should note that males of this species fly at great heights, at the very top of the trees, and descend very rarely to the ground [MIC 11, DEV 10, PEN 12, CHA 16]. It is therefore tempting to suppose that selective pressures that would favor iridescent blue individuals would no longer be relevant above the canopy or would be counterbalanced by deleterious effects (see section 9.4.3 for discussion on luminous flashes emitted by escape flight, probably less effective in gliding scenarios). Moreover, a number of ecological parameters change with height – such as habitat openness, increased sunlight or the nature and number of predators – which can influence the selective value of a phenotype. Three species said to be "canopy butterflies" have, however, intense and iridescent blue color (*M. cypris*, *M. rhetenor* and *M. anaxibia*), which shows that the canopy/understory dichotomy cannot exclusively explain the presence or absence of blue.

In the case of the three white *Morpho* species, who do not belong to one sole clade, their close resemblance suggests some sort of convergence, but the ecological factors involved are unknown; they are unlikely to be linked to flight height since *M. polyphemus* flies high, even at the canopy level [YOU 72]. As regards the two palest species of the group, *"sulkowskyi"* (*M. sulkowskyi* and *M. lympharis*), males have wings with virtually no melanin. They fly in fairly open environments and often above the vegetation, displaying a pale yellow color which, at least theoretically, should make them rather inconspicuous [GOM 07]. These two species are found at high altitudes, most often approximately 2000–2500 m. This association between loss of melanin and high altitude is quite surprising, given the importance of melanin in thermoregulation, at least in temperate climates, where species generally have more melanin when they live at higher altitudes (e.g. [KIN 85]).

Finally, in some species, females show no blue coloration: as previously discussed, this situation could result from a difference of ecological context between sexes, females being selected for their cryptic colors (Wallace's hypothesis). It has been suggested that the ocher-orange colors associated with the brown-black patterning in the females of certain species (*M. rhetenor*, *M. cypris* and *M. aega*) could indicate cases of Batesian mimicry, where the models to be imitated may be toxic species from the *Danainae* family [GAY 16]. However, the mimicry is not very accurate, and *M. rhetenor* and *M. cypris* females are much larger than their supposed models, which would contradict this hypothesis.

9.5. Conclusions and open questions

The resounding message of this review is that despite their iconic status, *Morpho* remain relatively unknown especially for the evolution of their coloration. While their phylogeny is well established and the nano-structural basis of their color fully identified, the ecological and genetic factors of these traits are almost entirely unexplored. The hypotheses commonly called upon to explain the evolution of the iridescent blue color, whether in terms of sexual selection or escape from predators, have never been explicitly tested. These gaps leave us with a number of open questions: are non-blue species subject to different selective pressures (thermoregulation, hydrophobia, predation, communication)? Do they fly more slowly or simply in a different way than their blue relatives? Are there, within each species, differences in flight behavior between sexes? In the hypothesis regarding the importance of blue as a sexual signal, do *Morpho* show evidence of particular visual receptors? Are they sensitive to polarized light? All of these questions require experiments to be carried out that would allow us to study flight modalities and numerous behavioral traits. But they point above all, despite more than a century of collecting, toward the lack of understanding of the ecology of *Morpho*.

9.6. Acknowledgments

We thank Jim Mallet for his advice on references and Peter Vukusik for the copy of one of his articles.

9.7. Bibliography

[ALL 08] ALLEN C.E., BELDADE P., ZWAAN B.J. *et al.*, "Differences in the selection response of serially repeated color pattern characters: standing variation, development, and evolution", *BMC Evolutionary Biology*, vol. 8, no. 1, p. 94, 2008.

[ALL 11] ALLEN C.E., ZWAAN B.J., BRAKEFIELD P.M., "Evolution of sexual dimorphism in the Lepidoptera", *Annual Review of Entomology*, vol. 56, pp. 445–464, 2011.

[BAG 07] BAGNARA J.T., FERNANDEZ P.J., FUJII R., "On the blue coloration of vertebrates", *Pigment Cell Research*, vol. 20, no. 1, pp. 14–26, 2007.

[BEL 02] BELDADE P., KOOPS K., BRAKEFIELD P.M., "Developmental constraints versus flexibility in morphological evolution", *Nature*, vol. 416, no. 6883, pp. 844–847, 2002.

[BÉN 16] BÉNÉLUZ F., "Stades premiers", in GAYMAN J.M. (ed.), *Les* Morpho. *Distribution, diversification, comportement*, Association des Lépidoptéristes de France, Paris, 2016.

[BER 03] BERTHIER S., CHARRON E., DA SILVA A., "Determination of the cuticle index of the scales of the iridescent butterfly *Morpho menelaus*", *Optics Communications*, vol. 228, no. 4, pp. 349–356, 2003.

[BER 06] BERTHIER S., CHARRON E., BOULENGUEZ J., "Morphological structure and optical properties of the wings of Morphidae", *Insect Science*, vol. 13, no. 2, pp. 145–158, 2006.

[BER 07] BERTHIER S., *Iridescences: the physical colors of insects*, Springer Science & Business Media, 2007.

[BER 10] BERTHIER S., *Photonique des Morphos*, Springer Science & Business Media, 2010.

[BLA 07] BLANDIN P., *The Systematics of the Genus Morpho, Fabricius, 1807*, Hillside Books, Canterbury, 2007.

[BLA 13] BLANDIN P., PURSER B., "Evolution and diversification of Neotropical butterflies: insights from the biogeography and phylogeny of the genus *Morpho* Fabricius, 1807 (Nymphalidae: Morphinae), with a review of the geodynamics of South America", *Tropical Lepidoptera Research*, vol. 23, no. 2, pp. 62–85, 2013.

[BRE 06] BREUKER C.J., DEBAT V., KLINGENBERG C.P., "Functional evo-devo", *Trends in Ecology & Evolution*, vol. 21, no. 9, pp. 488–492, 2006.

[BRO 95] BROWER A.V., "Locomotor mimicry in butterflies? A critical review of the evidence", *Philosophical Transactions of the Royal Society of London B: Biological Sciences*, vol. 347, no. 1322, pp. 413–425, 1995.

[BUL 04] BULINA M.E., LUKYANOV K.A., YAMPOLSKY I.V. *et al.*, "New class of blue animal pigments based on Frizzled and Kringle protein domains", *Journal of Biological Chemistry*, vol. 279, no. 42, pp. 43367–43370, 2004.

[CAS 10] CASSILDÉ C., BLANDIN P., PIERRE J. *et al.*, "Phylogeny of the genus *Morpho* Fabricius, 1807, revisited (Lepidoptera, Nymphalidae)", *Bulletin de la Société entomologique de France*, vol. 115, no. 2, 2010.

[CAS 12] CASSILDÉ C., BLANDIN P., SILVAIN J.F., "Phylogeny of the genus *Morpho* Fabricius 1807: insights from two mitochondrial genes (Lepidoptera: Nymphalidae)", *Annales de la Société entomologique de France*, vol. 48, nos 1–2, pp. 173–188, Taylor & Francis Group, January 2012.

[CHA 88] CHAI P., "Wing coloration of free-flying Neotropical butterflies as a signal learned by a specialized avian predator", *Biotropica*, pp. 20–30, 1988.

[CHA 16] CHAZOT N., PANARA S., ZILBERMANN N. *et al.*, "*Morpho* morphometrics: Shared ancestry and selection drive the evolution of wing size and shape in *Morpho* butterflies", *Evolution*, vol. 70, no. 1, pp. 181–194, 2016.

[CHO 13] CHO E.H., NIJHOUT H.F., "Development of polyploidy of scale-building cells in the wings of Manduca sexta", *Arthropod Structure & Development*, vol. 42, no. 1, pp. 37–46, 2013.

[CLE 66] CLENCH H.K., "Behavioral thermoregulation in butterflies", *Ecology*, vol. 47, no. 6, pp. 1021–1034, 1966.

[COS 07] COSTANZO K., MONTEIRO A., "The use of chemical and visual cues in female choice in the butterfly Bicyclus anynana", *Proceedings of the Royal Society of London B: Biological Sciences*, vol. 274, pp. 845–851, 2007.

[DAB 84] D'ABRERA B., *Butterflies of the Neotropical Region. Part II. Danaidae, Ithomiidae, Heliconidae & Morphidae*, Hill House, Ferny Creek, 1984.

[DEV 85] DEVRIES P.J., KITCHING I.J., VANE-WRIGHT R.I., "The systematic position of Antirrhea and Caerois, with comments on the classification of the Nymphalidae (Lepidoptera)", *Systematic Entomology*, vol. 10, no. 1, pp. 11–32, 1985.

[DEV 10] DEVRIES P.J., PENZ C.M., HILL R.I., "Vertical distribution, flight behaviour and evolution of wing *morphology* in *Morpho* butterflies", *Journal of Animal Ecology*, vol. 79, no. 5, pp. 1077–1085, 2010.

[DIN 14] DINWIDDIE A., NULL R., PIZZANO M. *et al.*, "Dynamics of F-actin prefigure the structure of butterfly wing scales", *Developmental Biology*, vol. 392, no. 2, pp. 404–418, 2014.

[DOU 07] DOUGLAS J.M., CRONIN T.W., CHIOU T.H. *et al.*, "Light habitats and the role of polarized iridescence in the sensory ecology of neotropical nymphalid butterflies (Lepidoptera: Nymphalidae)", *Journal of Experimental Biology*, vol. 210, no. 5, pp. 788–799, 2007.

[DOU 09] DOUCET S.M., MEADOWS M.G., "Iridescence: a functional perspective", *Journal of The Royal Society Interface*, vol. 6, no. 2, pp. S115–S132, 2009.

[END 88] ENDLER J.A., GREENWOOD J.J.D., "Frequency-dependent predation, crypsis and aposematic coloration", *Philosophical Transactions of the Royal Society of London B: Biological Sciences*, vol. 319, no. 1196, pp. 505–523, 1988.

[END 78] ENDLER J.A., "A predator's view of animal color patterns", *Evolutionary Biology*, pp. 319–364, Springer US, 1978.

[FRE 04] FREITAS A.V.L., BROWN K.S., "Phylogeny of the Nymphalidae (Lepidoptera)", *Systematic biology*, vol. 53, no. 3, pp. 363–383, 2004.

[FRU 12] FRUHSTORFER H., *Familie: Morphidae West*, in SEITZ A. (ed.), *Die Gross-Schmetterlinge der Erde*, vol. 5, Alfred Kernen, Stuttgart, 1912.

[GAL 98] GALANT R., SKEATH J.B., PADDOCK S. *et al.*, "Expression pattern of a butterfly achaete-scute homolog reveals the homology of butterfly wing scales and insect sensory bristles", *Current Biology*, vol. 8, no. 14, pp. 807–813, 1998.

[GAR 94] GARLAND JR T., LOSOS J.B., "Ecological morphology of locomotor performance in squamate reptiles", *Ecological morphology: integrative organismal biology*, pp. 240–302, 1994.

[GAR 09] GARCÍA-BELLIDO A., DE CELIS J.F., "The complex tale of the achaete–scute complex: a paradigmatic case in the analysis of gene organization and function during development", *Genetics*, vol. 182, no. 3, pp. 631–639, 2009.

[GAY 16] GAYMAN J.-M., MERLIER F., OUVAROFF J. *et al.*, *Les Morpho. Distribution, diversification, comportement*, Association des Lépidoptéristes de France, 2016.

[GHI 72] GHIRADELLA H., ANESHANSLEY D., EISNER T. *et al.*, "Ultraviolet reflection of a male butterfly: interference color caused by thin-layer elaboration of wing scales", *Science*, vol. 178, no. 4066, pp. 1214–1217, 1972.

[GHI 76] GHIRADELLA H., RADIGAN W., "Development of butterfly scales. II. Struts, lattices and surface tension", *Journal of Morphology*, vol. 150, no. 2, pp. 279–297, 1976.

[GHI 89] GHIRADELLA H., "Structure and development of iridescent butterfly scales: lattices and laminae", *Journal of Morphology*, vol. 202, no. 1, pp. 69–88, 1989.

[GIL 88] GILBERT L.E., FORREST H.S., SCHULTZ T.D. *et al.*, "Correlations of ultrastructure and pigmentation suggest how genes control development of wing scales of Heliconius butterflies", *The Journal of Research on the Lepidoptera*, vol. 26, pp. 141–160, 1988.

[GIR 16] GIRALDO M.A., YOSHIOKA S., LIU C. *et al.*, "Coloration mechanisms and phylogeny of *Morpho* butterflies", *Journal of Experimental Biology*, vol. 219, no. 24, pp. 3936–3944, 2016.

[GOM 07] GOMEZ D., THÉRY M., "Simultaneous crypsis and conspicuousness in color patterns: comparative analysis of a neotropical rainforest bird community", *The American Naturalist*, vol. 169, no. S1, pp. S42–S61, 2007.

[GOU 02] GOULD S.J., *The Structure of Evolutionary Theory*, Harvard University Press, 2002.

[HEL 12] HELICONIUS GENOME CONSORTIUM, "Butterfly genome reveals promiscuous exchange of mimicry adaptations among species", *Nature*, vol. 487, pp. 94–98, doi: 10.1038/nature11041, 2012.

[HIL 11] HILL G.E., "Condition-dependent traits as signals of the functionality of vital cellular processes", *Ecology Letters*, vol. 14, no. 7, pp. 625–634, 2011.

[HIN 73] HINTON H.E., "Natural deception", in *Illusion in nature and art*, Duckworth, London, 1973.

[ING 08] INGRAM A.L., PARKER A.R., "A review of the diversity and evolution of photonic structures in butterflies, incorporating the work of John Huxley (The Natural History Museum, London from 1961 to 1990)", *Philosophical Transactions of the Royal Society of London B: Biological Sciences*, vol. 363, no. 1502, pp. 2465–2480, 2008.

[JOR 11] JORON M., FREZAL L., JONES R.T. *et al.*, "Chromosomal rearrangements maintain a polymorphic supergene controlling butterfly mimicry", *Nature*, vol. 477, pp. 203–206, doi: 10.1038/nature10341, 2011.

[KEM 07] KEMP D.J., "Female butterflies prefer males bearing bright iridescent ornamentation", *Proceedings of the Royal Society of London B: Biological Sciences*, vol. 274, pp. 1043–1047, 2007.

[KEM 07] KEMP D.J., "Female mating biases for bright UV iridescence in a butterfly *Eurema hecabe* (Pieridae)", *Behavioral Ecology*, vol. 19, pp. 1–8, 2007.

[KEM 11] KEMP D.J., RUTOWSKI R.L., "The role of coloration in mate choice and sexual interactions in butterflies", *Advances in the Study of Behavior*, vol. 43, no. 5, 2011.

[KEM 14] KEMP D.J., JONES D., MACEDONIA J.M. *et al.*, "Female mating preferences and male signal variation in iridescent Hypolimnas butterflies", *Animal Behaviour*, vol. 87, pp. 221–229, 2014.

[KIN 85] KINGSOLVER J.G., "Thermoregulatory significance of wing melanization in Pieris butterflies (Lepidoptera: Pieridae): physics, posture, and pattern", *Oecologia*, vol. 66, no. 4, pp. 546–553, 1985.

[KIN 02] KINOSHITA S., YOSHIOKA S., KAWAGOE K., "Mechanisms of structural colour in the *Morpho* butterfly: cooperation of regularity and irregularity in an iridescent scale", *Proceedings of the Royal Society of London B: Biological Sciences*, vol. 269, no. 1499, pp. 1417–1421, 2002.

[KOO 00] KOON D.W., CRAWFORD A.B., "Insect thin films as sun blocks, not solar collectors", *Applied Optics*, vol. 39, no. 15, pp. 2496–2498, 2000.

[KOT 80] KOTTLER M.J., "Darwin, Wallace, and the origin of sexual dimorphism", *Proceedings of the American Philosophical Society*, vol. 124, no. 3, pp. 203–226, 1980.

[KUM 94] KUMAZAWA K., TANAKA S., NEGITA K. *et al.*, "Fluorescence from wing of *Morpho sulkowskyi* butterfly", *Japanese Journal of Applied Physics*, vol. 33, no. 4R, p. 2119, 1994.

[LAG 15] LAGORIO M.G., CORDON G.B., IRIEL A. *et al.*, "Reviewing the relevance of fluorescence in biological systems", *Photochemical & Photobiological Sciences*, vol. 14, no. 9, pp. 1538–1559, 2015.

[LAM 04] LAMAS G., "Tribe *Morphini*", in HEPPNER J.B. (ed.), *Checklist: Part 4a, Hesperoidea - Papilionoidea – Atlas of Neotropical Lepidoptera*, vol. 5a, Association for Neotropical Lepidoptera, Scientific Publishers, Gainesville, 2004.

[LE 14] LE POUL Y., WHIBLEY A., CHOUTEAU M. *et al.*, "Evolution of dominance mechanisms at a butterfly mimicry supergene", *Nature Communications*, vol. 5, p. 5644, doi:10.1038/ncomms6644, 2014.

[MAL 87] MALLET J., SINGER M.C., "Individual selection, kin selection, and the shifting balance in the evolution of warning colours: the evidence from butterflies", *Biological Journal of the Linnean Society*, vol. 32, no. 4, pp. 337–350, 1987.

[MAL 99] MALLET J., JORON M., "Evolution of diversity in warning color and mimicry: polymorphisms, shifting balance, and speciation", *Annual Review of Ecology and Systematics*, vol. 30, no. 1, pp. 201–233, 1999.

[MEJ 13] MEJDOUBI A., ANDRAUD C., BERTHIER S. *et al.*, "Finite element modeling of the radiative properties of *Morpho* butterfly wing scales", *Physical Review E*, vol. 87, no. 2, p. 022705, 2013.

[MIC 11] MICHAEL O., "Lebensweise und Gewohnheiten der *Morpho* des Amazonasgebietes", *Fauna Exotica*, vol. 1, pp. 10–17, 1911.

[NAD 16] NADEAU N.J., PARDO-DIAZ C., WHIBLEY A. *et al.*, "The gene cortex controls mimicry and crypsis in butterflies and moths", *Nature*, vol. 534, pp. 106–110, doi:10.1038/nature17961, 2016.

[NAT 16] NATTIER R., CAPDEVIELLE-DULAC C., CASSILDÉ C. *et al.*, "Phylogeny and diversification of the cloud forest *Morpho* sulkowskyi group (Lepidoptera, Nymphalidae) in the evolving Andes", *Zoologica Scripta*, 2016.

[NEI 08] NEILD A.F.E., *The Butterflies of Venezuela. Part 2: Nymphalidae II (Acraeinae, Libytheinae, Nymphalinae, Ithomiinae, Morphinae)*, Meridian Publications, London, 2008.

[NIJ 91] NIJHOUT H.F., *The development and evolution of butterfly wing patterns*, Smithsonian, Institution Scholarly Press, 1991.

[OLS 98] OLSON V.A., OWENS I.P., "Costly sexual signals: are carotenoids rare, risky or required?", *Trends in Ecology & Evolution*, vol. 13, no. 12, pp. 510–514, 1998.

[OVE 66] OVERTON J., "Microtubules and microfibrils in morphogenesis of the scale cells of Ephestia kühniella", *The Journal of Cell Biology*, vol. 29, no. 2, pp. 293–305, 1966.

[PEN 02] PENZ C.M., DEVRIES P.J., "Phylogenetic analysis of *Morpho* butterflies (Nymphalidae, Morphinae): implications for classification and natural history", *American Museum Novitates*, pp. 1–33, 2002.

[PEN 06] PEÑA C., WAHLBERG N., WEINGARTNER E. *et al.*, "Higher level phylogeny of Satyrinae butterflies (Lepidoptera: Nymphalidae) based on DNA sequence data", *Molecular Phylogenetics and Evolution*, vol. 40, no. 1, pp. 29–49, 2006.

[PEN 12] PENZ C.M., DEVRIES P.J., WAHLBERG N., "Diversification of *Morpho* butterflies (Lepidoptera, Nymphalidae): a re-evaluation of *morpho*logical characters and new insight from DNA sequence data", *Systematic Entomology*, vol. 37, no. 4, pp. 670–685, 2012.

[PIK 15] PIKE T.W., "Interference coloration as an anti-predator defence", *Biology Letters*, vol. 11, no. 4, p. 20150159, 2015.

[PIN 96] PINHEIRO C.E., "Palatablility and escaping ability in Neotropical butterflies: tests with wild kingbirds (Tyrannus melancholicus, Tyrannidae)", *Biological Journal of the Linnean Society*, vol. 59, no. 4, pp. 351–365, 1996.

[PIN 16] PINHEIRO C.E.G., FREITAS A.V.L., CAMPOS V.C. *et al.*, "Both palatable and unpalatable butterflies use bright colors to signal difficulty of capture to predators", *Neotropical Entomology*, vol. 45, no. 2, pp. 107–113, 2016.

[POT 07] POTYRAILO R.A., GHIRADELLA H., VERTIATCHIKH A. *et al.*, "*Morpho* butterfly wing scales demonstrate highly selective vapour response", *Nature Photonics*, vol. 1, no. 2, pp. 123–128, 2007.

[PRU 15] PRUDIC K.L., STOEHR A.M., WASIK B.R. *et al.*, "Eyespots deflect predator attack increasing fitness and promoting the evolution of phenotypic plasticity", in *Proceedings of the Royal Society of London B: Biological Sciences*, vol. 282, no. 1798, p. 20141531, January 2015.

[RAM 14] RAMÍREZ GARCIA C., GALLUSSER S., LACHAUME G. *et al.*, "The ecology and life cycle of the Amazonian *Morpho cisseis phanodemus* Hewitson, 1869, with a comparative review of early stages in the genus *Morpho* (Lepidoptera: Nymphalidae: Morphinae)", *Tropical Lepidoptera Research*, vol. 24, no. 2, pp. 67–80, 2014.

[REV 12] REVELL L.J., "Phytools: an R package for phylogenetic comparative biology (and other things)", *Methods in Ecology and Evolution*, vol. 3, no. 2, pp. 217–223, 2012.

[ROB 69] ROBINSON M.H., "Defenses against visually hunting predators", *Journal of Evolutionary Biology*, vol. 3, no. 22, pp. 5–59, 1969.

[ROB 05] ROBERTSON K.A., MONTEIRO A., "Female Bicyclus anynana butterflies choose males on the basis of their dorsal UV-reflective eyespot pupils", *Proceedings of the Royal Society of London B: Biological Sciences*, vol. 272, no. 1572, pp. 1541–1546, 2005.

[RUX 04] RUXTON G.D., SPEED M., SHERRATT T.N., "Evasive mimicry: when (if ever) could mimicry based on difficulty of capture evolve?", *Proceedings of the Royal Society of London B: Biological Sciences*, vol. 271, no. 1553, pp. 2135–2142, 2004.

[SAR 10] SARANATHAN V., OSUJI C.O., MOCHRIE S.G. *et al.*, "Structure, function, and self-assembly of single network gyroid (I4132) photonic crystals in butterfly wing scales", *Proceedings of the National Academy of Sciences*, vol. 107, no. 26, pp. 11676–11681, 2010.

[SCH 12] SCHNEIDER C.A., RASBAND W.S., ELICEIRI K.W., "NIH Image to ImageJ: 25 years of image analysis", *Nature Methods*, vol. 9, pp. 671–675, 2012.

[SID 13] SIDDIQUE R.H., DIEWALD S., LEUTHOLD J. *et al.*, "Theoretical and experimental analysis of the structural pattern responsible for the iridescence of *Morpho* butterflies", *Optics Express*, vol. 21, no. 12, pp. 14351–14361, 2013.

[SIL 84] SILBERGLIED R.E., "Visual communication and sexual selection among butterflies", in *Symposia of the Royal Entomological Society of London*, 1984.

[SIM 90] SIMPSON P., "Lateral inhibition and the development of the sensory bristles of the adult peripheral nervous system of Drosophila", *Development*, vol. 109, no. 3, pp. 509–519, 1990.

[SKE 91] SKEATH J.B., CARROLL S.B., "Regulation of achaete-scute gene expression and sensory organ pattern formation in the Drosophila wing", *Genes & Development*, vol. 5, no. 6, pp. 984–995, 1991.

[STE 07] STEVENS M., "Predator perception and the interrelation between different forms of protective coloration", *Proceedings of the Royal Society of London B: Biological Sciences*, vol. 274, no. 1617, pp. 1457–1464, 2007.

[TAD 99] TADA H., MANN S.E., MIAOULIS I.N. *et al.*, "Effects of a butterfly scale microstructure on the iridescent color observed at different angles", *Optics Express*, vol. 5, no. 4, pp. 87–92, 1999.

[UMB 13] UMBERS K.D., "On the perception, production and function of blue colouration in animals", *Journal of Zoology*, vol. 289, no. 4, pp. 229–242, 2013.

[VAN 59] VAN SOMEREN V.G.L., JACKSON T.H.E., "Some comments on protective resemblance amongst African Lepidoptera (Rhopalocera)", *Journal of the Lepidopterists Society*, vol. 13, pp. 121–150, 1959.

[VAN 11a] VAN HOOIJDONK E., BARTHOU C., VIGNERON J.P. *et al.*, "Detailed experimental analysis of the structural fluorescence in the butterfly *Morpho sulkowskyi* (Nymphalidae)", *Journal of Nanophotonics*, vol. 5, no. 1, pp. 053525–053525, 2011.

[VAN 11b] VAN HOOIJDONK E., BARTHOU C., VIGNERON J.P. *et al.*, "Structural iridescence in the butterfly *Morpho sulkowskyi* (Nymphalidae)", *Proceedings of SPIE*, vol. 80940, 2011.

[VUK 99] VUKUSIC P., SAMBLES J.R., LAWRENCE C.R. *et al.*, "Quantified interference and diffraction in single *Morpho* butterfly scales", *Proceedings of the Royal Society of London B: Biological Sciences*, vol. 266, no. 1427, pp. 1403–1411, 1999.

[VUK 00] VUKUSIC P., SAMBLES J.R., GHIRADELLA H., "Optical classification of microstructure in butterfly wing-scales", *Photonics Science News*, vol. 6, no. 1, pp. 61–66, 2000.

[VUK 03] VUKUSIC P., SAMBLES J.R., "Photonic structures in biology", *Nature*, vol. 424, no. 6950, pp. 852–855, 2003.

[WAG 96] WAGNER T., NEINHUIS C., BARTHLOTT W., "Wettability and contaminability of insect wings as a function of their surface sculptures", *Acta Zoologica*, vol. 77, no. 3, pp. 213–225, 1996.

[WAS 75] WASSERTHAL L.T., "The role of butterfly wings in regulation of body temperature", *Journal of Insect Physiology*, vol. 21, no. 12, pp. 1921–1930, 1975.

[WAS 14] WASIK B.R., LIEW S.F., LILIEN D.A. *et al.*, "Artificial selection for structural color on butterfly wings and comparison with natural evolution", *Proceedings of the National Academy of Sciences*, vol. 111, no. 33, pp. 12109–12114, 2014.

[WIL 11] WILTS B.D., PIRIH P., STAVENGA D.G., "Spectral reflectance properties of iridescent pierid butterfly wings", *Journal of Comparative Physiology A*, vol. 197, no. 6, pp. 693–702, 2011.

[WIL 16] WILTS B.D., GIRALDO M.A., STAVENGA D.G., "Unique wing scale photonics of male Rajah Brooke's birdwing butterflies", *Frontiers in Zoology*, vol. 13, no. 1, p. 36, 2016.

[YOS 04] YOSHIOKA S., KINOSHITA S., "Wavelength–selective and anisotropic light–diffusing scale on the wing of the *Morpho* butterfly", *Proceedings of the Royal Society of London B: Biological Sciences*, vol. 271, no. 1539, pp. 581–587, 2004.

[YOS 06] YOSHIOKA S., KINOSHITA S., "Structural or pigmentary? Origin of the distinctive white stripe on the blue wing of a *Morpho* butterfly", *Proceedings of the Royal Society of London B: Biological Sciences*, vol. 273, no. 1583, pp. 129–134, 2006.

[YOU 71] YOUNG A.M., "Wing coloration and reflectance in *Morpho* butterflies as related to reproductive behavior and escape from avian predators", *Oecologia*, vol. 7, no. 3, pp. 209–222, 1971.

[YOU 72] YOUNG A.M., MUYSHONDT A., "Biology of *Morpho* polyphemus (Lepidoptera: Morphidae) in El Salvador", *Journal of the New York Entomological Society*, pp. 18–42, 1972.

[YOU 73] YOUNG A.M., "Studies on comparative ecology and ethology in adult populations of several species of *Morpho* butterflies (Lepidoptera: Morphidae)", *Studies on Neotropical Fauna and Environment*, vol. 8, no. 1, pp. 17–50, 1973.

[WAH 09] WAHLBERG N., LENEVEU J., KODANDARAMAIAH U. *et al.*, "Nymphalid butterflies diversify following near demise at the Cretaceous/Tertiary boundary", *Proceedings of the Royal Society of London B: Biological Sciences*, vol. 276, no. 1677, pp. 4295–4302, 2009.

[ZHE 07] ZHENG Y., GAO X., JIANG L., "Directional adhesion of superhydrophobic butterfly wings", *Soft Matter*, vol. 3, no. 2, pp. 178–182, 2007.

Biodiversity in Natural History Collections: a Source of Data for the Study of Evolution

10.1. Introduction

Natural history collections (NHC), most often kept in research institutions such as museums or universities, are spatio-temporal testimonies to biological diversity. Indeed, they provide information about the presence of species in a given place, during a given period. They therefore represent what we currently know about biodiversity [LAN 96]. It is however important to note that the historic role of the documentation of biodiversity depends on the wealth of these collections themselves. This wealth must be evaluated according to several criteria, such as the number of species and supra-specific taxons, a large diversity of source-locations, and also the number of specimens per species, reflecting intra-specific diversity.

NHCs play several important roles:

i) They create a space where samples of current and past biodiversity on Earth have been preserved for several centuries [LIS 11]. These collections therefore make it possible to document present and past diversity, including that of extinct species or fossils, and to study this diversity. They are therefore an important source of data for long-term studies allowing us to understand the influence of humankind on the environment [JOH 11, LIS 11].

Chapter written by Romain NATTIER.

ii) They also have the role of raising scientific awareness in the public by means of exhibitions, and inform educational programs intended for the public as well as students [LEM 81, ALL 94].

iii) They are an indispensable form of teaching support for university courses, whether that be in general biology (knowledge of the main groups) or in taxonomy and identification of less easily recognizable groups.

iv) They are a building block of expertise in this field, allowing us to identify specimens sent in by researchers, non-professionals or the public [ALL 94], and in addition, comparison with specimens serves as a basis for the description of species (specimen types).

When biodiversity and its future evolution are the topic of so much speculation, it is important to realize that an understanding of current biodiversity hinges on an understanding of its origins, the way in which it has been structured, and in which the species creating this biodiversity have diversified over time.

Two major obstacles limit the study of such historical processes from the basis of current data. The first, spatial in nature [LOM 04], corresponds to the difficulty in obtaining biological material from all the regions in which species are distributed, whether due to lack of means to sample often immense areas, to the difficulty in accessing zones of conflict, or to the growing difficulty of obtaining sampling permits. The second obstacle has to do with time [HAB 14]: the use of recent or living specimens only allows access to current morphological or genetic data, not to this sort of data from the past.

The significant biodiversity in collections makes it possible to bypass these obstacles by taking into account not only present biodiversity, but past biodiversity as well. NHCs are therefore remarkable tools for understanding the origin and evolution of biodiversity, but they remain underused in comparison to their genuine wealth, especially on the level of genetic resources [WAN 07, YEA 16]. The main obstacle to the molecular use of collection specimens is a technical one. The degradation of specimen DNA over time, starting with placement in a collection, leads to poor quality and quantity of useable DNA in older specimens. However, recently developed Next-Generation Sequencing (NGS) now makes it possible to sequence highly fragmented DNA and to further exploit often old collection specimens. In addition, there are several non-invasive protocols, specifically in the case of Arthropods, making it possible to extract DNA from collection specimens without causing even the slightest external damage to said specimens [GIL 07, THO 09]. These methodological advances make it possible to make the most of collections through new studies on the structuration and evolution of biological diversity.

In this article, I plan to present the advantages of collections as sources of molecular data for the study of evolution, and how the study of this biodiversity, represented by specimens, allows us to gain a better understanding of evolutionary processes.

10.2. Description of biodiversity

10.2.1. *Identification, comparison with specimen types*

The main role of collections is to preserve specimens, allowing us to describe species (specimen types). In this respect, comparison between unidentified specimens and types makes it possible either to specify their identification, or to describe new species. It is then possible in certain cases to test the link between the conserved specimen type in collections and a whole range of species difficult to determine from a morphological point of view [PRI 15]. For example, The Chomicki & Renner team [CHO 15] compared the typical specimen of watermelons (*Citrullus lanatus*), prepared by one of Linné's collectors (Thunberg) in 1773 in South Africa, with a variety of African species. The morphological characters within this genus are difficult to study, because of the great variability of leaves and the poor preservation of their flowers. The use of molecular tools has allowed authors to discover that the specimen type of *Citrullus lanatus* did not belong to a watermelon species but was rather a representative of the closely related endemic species of South Africa. This result has allowed authors to suggest that the origin of cultivated watermelon species is more in West Africa than South Africa.

10.2.2. *Phylogenetic position*

The use of NHCs also makes it possible to specify the phylogenetic position of extinct, or little known species, but whose specimens have been carefully conserved [PAR 04, ASH 06]. Wallander & Albert [WAL 00], using such a method, studied the monospecific genus *Hesperelaea* (Oleaceae), of which the only known specimen was collected by Edward

Palmer and described by Asa Gray in 1876 as *H. palmeri* on Guadalupe Island (Mexico). Upon collection, Palmer had only found three old trees still alive (no young trees and many dead) in a zone highly grazed on by sheep. Very little information is available on this genus: Gray's description is short, the floral morphology seen is unusual for an Oleaceae and no fossil of this genus is known, which makes comparison with other Oleaceae species, and thus its taxonomic positioning, very difficult. Sequencing of the specimen type has allowed us to specify its belonging to the Oleaceae and its phylogenetic positioning within this family.

10.2.3. *Delimitation of species*

Many methods of delimiting species using molecular data have been developed during recent years [FUJ 12, PUI 12]. These methods compare molecular data coming from significant series of specimens in order to distinguish the different species therein. As complete a taxonomic overview of the studied group as possible is necessary to take into account the variability of different molecular markers used and so it is that NHCs provide the necessary material for these studies. In combination with morphological and geographic data, NHC specimens thus allow us to determine different species' delimitation among themselves [SMI 08, LEA 10, MAD 14]. In some cases this can even lead to the description of new species conserved for several decades in NHCs, until then unknown [BEB 10]. In this sense, the molecular analysis of material in collections makes it possible to accelerate the rate of discovery of new species [YEA 16] and to better understand the structuration of biodiversity, a prerequisite for studying ecological and evolutionary processes.

10.3. Ecological and evolutionary processes on a population level

The evolution of the genetic structure of a species depends on both historical and present factors. As a consequence, in order to determine the processes behind this evolution, it is necessary to study a chronological series of specimens [HAB 14]. If the studied specimens only come from recent collections, this results in an incomplete vision of the biological processes at work. The study of specimens collected before and after the process or event under study allows us to specify its impact on populations

[HAB 14]. In this light, collections prove to be a unique source of resources to understand how species respond to changes in their environment, notably those collections with an intra-specific sample range, which is important on a spatial and temporal scale. This sample range allows us to study variations in genetic diversity. This parameter is important as it is generally correlated with a higher capacity of adaptation and a smaller risk of extinction for a species or population. For example, collections collated over the past two centuries allow us to specify the response of organisms to the intensification of man-made pressures by facilitating the study of genetic diversity before and during this intensification [WAN 07, BI 13].

In the case of biological invasions, the first phases, which are the introduction of the exotic population and its establishment in a new territory, are often misunderstood. Collections are therefore able to provide a historical dimension to this invasion process, if collections have been regularly carried out in the zone under study [SUA 05, CRA 09], of which often only the current geographical dimension is known. During this initial phase, an invasive population is not yet recognized as such and it is possible to study its genetic and phenotypic variability before its phase of expansion [HAR 06b]. Invasive populations are often over-represented in NHCs during this phase due to collection bias in favor of rare or new species, which makes studying them easier. The two key factors in understanding the invasion, namely its rate and the source of the introduced species, allow us to study the invasion dynamic and can be informed by the study of NHCs [RUS 08]. In a similar vein of thought, the study of NHCs also allows us to model the ecological niches of invasive species, with a view to proposing predictive models regarding future invasions [WAR 07].

NHCs have also made it possible to specify the influence of modification and fragmentation of habitats, notably on the variation in genetic diversity. One of the first studies carried out in this area showed a loss of alleles following the fragmentation of a habitat in pinnated grouse (*Tympanuchus cupido*) in the USA [BOU 98]. Other studies, both in mammals [MIL 03] and insects [HAR 06a, HAB 11, UGE 11], have also demonstrated this process with the help of chronological specimen series. In the case of pesticidal resistance, Délye's team [DEL 13] studied slender meadow foxtail (*Alopecurus myosuroides*) in order to determine the frequency of mutants before the use of a herbicide. The authors used collection specimens in order to search for mutants from among more than 700 specimens predating the use of herbicides since 1788. They thus highlighted the presence of mutants of this type before the use of herbicides: this mutation may therefore be present with no major negative effect and contribute to their genetic variation. In this case, the use of herbicides nonetheless accelerated resistance in the population by selecting individuals carrying these mutations. Other studies have allowed us to test the effects of climate change, with a modification of phenology in insects [BAR 11, POL 13] and in mushrooms [KAU 08], a change of coloration in birds living in snowy regions [KAR 11], of body size in birds [GAR 14a] and mammals [MEI 09].

10.4. Ecological and evolutionary processes on a phylogenetic scale

By virtue of their geographical distribution or certain anatomical characteristics, certain extinct species are considered as key elements to answer evolutionary questions. The use of NGS methods has made it possible to amplify DNA fragments from old specimens, which has made it easier to do molecular analysis on collection specimens, and allows for a more significant sample range of lines of descent that were previously difficult or indeed impossible to incorporate into phylogenetic studies. This aspect is all the more important for the numerous molecular methods of biodiversity analysis (measures of the diversification of lines of descent, for example), which require as complete a taxonomic overview as possible.

Collection specimens notably allow us to study the biogeographic origin of extinct species. This is the case, for example, for *Sicyos villosus* (Cucurbitaceae), a species from the Galapagos islands. The only specimen of this species was collected in a herbarium by Charles Darwin in September 1835, and the species has not been seen since and is thus considered to be extinct. Sebastian's team [SEB 10] (2010) was able to extract DNA from one of the roots of the specimen conserved in Darwin's herbarium and sequence several fragments of the DNA. Their analysis shows that this species is in fact closely related to other species in North America and Mexico, and that it may have arrived on the island via distribution from a continental source up to 3–4 thousand years ago. Potential scenarios of dispersal can also be tested with more reliability, for example, in the case of endemic rodents from the Caribbean [FAB 14]. The evolutionary study of endangered species is made complex by the difficulty in collecting them in the field and by their low representation in captivity. This is the case, for example, with Guenons (Primates), of which the biogeographic origin and the rate of their diversification have been reconstructed solely from collection specimens [GUS 13]. The evolutionary history of emblematic extinct species can also be traced by specimens from NHCs. It is, for example, the case for the dodo (*Raphus cucullatus*), endemic to the Mauritius islands and extinct at the end of the 17th Century. Despite the numerous morphological differences from closely related species, the Shapiro team [SHA 02] showed that the dodo was part of the Columbidae family, not a separate family in its own right, among other small wingless species endemic to the islands. This study made it possible to conclude that the ancestors of the dodo spread from South East Asia around 26 million years ago.

10.5. Conclusion

As shown by all these examples, NHCs are important sources of data for numerous studies on biodiversity. While NHC specimens are being used more and more to document the impact of global changes on biodiversity [PYK 10], funding allocated to the management, enrichment and development of these collections is constantly being reduced in many countries [DAL 03, FRO 03, AND 14, GAR 14b]. The significant biodiversity within collections is nonetheless a source of data that allows us to better understand evolutionary mechanisms, and therefore to predict the effects of changes that are primarily man-made. Collection and conservation of the species most common in NHCs are the most heavily affected, while various studies have highlighted the importance of conserving chronological sample series of this morphological and genetic diversity [SCH 15]. How can we add to collections to make them useful for future generations? One of the ways in which we can continue to collect these series and to add specimens to museums would be to return to the various different roles of collections, by opening them up to a larger public. This would work by encouraging the public to gain a better understanding of collection practices and of the role of collections, but also by encouraging biology teachers, both at secondary school and at universities, to integrate these collections into their teaching syllabus [REN 16].

It is of the utmost importance to safeguard specimens in these NHCs for future studies, and to document as much information as possible (photos, dates, habitats, methods of capture, GPS, DNA) [KRE 14], essential to future evolutionary studies. Finally, even if recent developments in NGS technology allow us to sequence very quickly and at a lower cost, old specimens, the taxonomy, the description of biodiversity and the store of specimen types in NHCs still provide an essential platform for evolutionary studies by providing the very objects of these studies.

For NHC specimens to be adequately valued, it is necessary, on the one hand, that scientists use the data available therein to resolve taxonomic and evolutionary problems of various scales, but also that citizens have a greater awareness of the biodiversity hidden in museums by playing a more active role in studying them. There already exist several citizen science programs that raise awareness of these collections (in the herbarium of the Muséum national d'Histoire naturelle, for example - http://lesherbonautes.mnhn.fr), and also make it possible to test evolutionary hypotheses in nature (for example, in grove snails - http://www.evolutionmegalab.org). On the other hand, in the future it would be necessary to organize a research program taking into consideration all of these aspects (citizen involvement, biodiversity, collection and evolution), thus allowing for a more integrative approach to getting the most out of NHCs.

10.6. Acknowledgements

I would like to thank Marie-Christine Maurel and Philippe Grandcolas, organizers of the ISYEB's scientific day "Evolution and Biodiversity", as well as Laure Turcati, Frédéric Legendre, Tony Robillard and Alice Michel-Salzat for their critical revision of this article.

10.7. Bibliography

[ALL 94] ALLMON W.D., "The Value of Natural History Collections", *Curator: The Museum Journal*, vol. 37, pp. 83–89, 1994.

[AND 14] ANDREONE F., BARTOLOZZI L., BOANO G. *et al.*, "Italian natural history museums on the verge of collapse?", *ZooKeys*, vol. 456, pp. 139–146, 2014.

[ASH 06] ASHER R.J., HOFREITER M., "Tenrec phylogeny and the noninvasive extraction of nuclear DNA", *Systematic Biology*, vol. 55, pp. 181–194, 2006.

[BAR 11] BARTOMEUS I., ASCHER J.S., WAGNER D. *et al.*, "Climate-associated phenological advances in bee pollinators and bee-pollinated plants", *Proceedings of the National Academy of Sciences*, vol. 108, pp. 20645–20649, 2011.

[BEB 10] BEBBER D.P., CARINE M.A., WOOD J.R.I. *et al.*, "Herbaria are a major frontier for species discovery", *Proceedings of the National Academy of Sciences*, vol. 107, pp. 22169–22171, 2010.

[BI 13] BI K., LINDEROTH T., VANDERPOOL D. *et al.*, "Unlocking the vault: next-generation museum population genomics", *Molecular Ecology*, vol. 22, pp. 6018–6032, 2013.

[BOU 98] BOUZAT J.L., LEWIN H.A., PAIGE K.N., "The ghost of genetic diversity past: historical DNA analysis of the greater prairie chicken", *The American Naturalist*, vol. 152, pp. 1–6, 1998.

[CHO 15] CHOMICKI G., RENNER S.S., "Watermelon origin solved with molecular phylogenetics including Linnaean material: another example of museomics", *New Phytologist*, vol. 205, pp. 526–532, 2015.

[CRA 09] CRAWFORD P.H.C., HOAGLAND B.W., "Can herbarium records be used to map alien species invasion and native species expansion over the past 100 years?", *Journal of Biogeography*, vol. 36, pp. 651–661, 2009.

[CUT 16] CUTLIP K., Smithsonian's mosquito collection is weapon in battle against Zika, available at: http://insider.si.edu/2016/06/smithsonians-mosquito-collection-weapon-battle-zika, 2016

[DAL 03] DALTON R., "Natural history collections in crisis as funding is slashed", *Nature*, vol. 423, pp. 575–575, 2003.

[DÉL 13] DÉLYE C., DEULVOT C., CHAUVEL B., "DNA analysis of herbarium specimens of the grass weed *Alopecurus myosuroides* reveals herbicide resistance pre-dated herbicides", *PLoS ONE*, vol. 8, p. e75117, 2013.

[FAB 14] FABRE P.-H., VILSTRUP J.T., RAGHAVAN M. *et al.*, "Rodents of the Caribbean: origin and diversification of hutias unravelled by next-generation museomics", *Biology Letters*, vol. 10, p. 20140266, 2014.

[FRO 03] FROELICH A., "Smithsonian science: first class on a coach budget", *BioScience*, vol. 53, p. 328, 2003.

[FUJ 12] FUJITA M.K., LEACHÉ A.D., BURBRINK F.T. *et al.*, "Coalescent-based species delimitation in an integrative taxonomy", *Trends in Ecology & Evolution*, vol. 27, pp. 480–488, 2012.

[GAR 14a] GARDNER J.L., AMANO T., BACKWELL P.R.Y. *et al.*, "Temporal patterns of avian body size reflect linear size responses to broadscale environmental change over the last 50 years", *Journal of Avian Biology*, vol. 45, pp. 529–535, 2014.

[GAR 14b] GARDNER J.L., AMANO T., SUTHERLAND W.J. *et al.*, "Are natural history collections coming to an end as time-series?: Peer-reviewed letter", *Frontiers in Ecology and the Environment*, vol. 12, pp. 436–438, 2014.

[GIL 07] GILBERT M.T.P., MOORE W., MELCHIOR L. *et al.*, "DNA extraction from dry museum beetles without conferring external morphological damage", *PLoS ONE*, vol. 2, p. e272, 2007.

[GUS 13] GUSCHANSKI K., KRAUSE J., SAWYER S. *et al.*, "Next-generation museomics disentangles one of the largest primate radiations", *Systematic Biology*, vol. 62, pp. 539–554, 2013.

[HAB 11] HABEL J.C., FINGER A., SCHMITT T. *et al.*, "Survival of the endangered butterfly *Lycaena helle* in a fragmented environment: Genetic analyses over 15 years", *Journal of Zoological Systematics and Evolutionary Research*, vol. 49, pp. 25–31, 2011.

[HAB 14] HABEL J.C., HUSEMANN M., FINGER A. *et al.*, "The relevance of time series in molecular ecology and conservation biology", *Biological Reviews*, vol. 89, pp. 484–492, 2014.

[HAR 06a] HARPER G.L., MACLEAN N., GOULSON D., "Analysis of museum specimens suggests extreme genetic drift in the adonis blue butterfly (*Polyommatus bellargus*)", *Biological Journal of the Linnean Society*, vol. 88, pp. 447–452, 2006.

[HAR 06b] HARTLEY S., HARRIS R., LESTER P.J., "Quantifying uncertainty in the potential distribution of an invasive species: climate and the Argentine ant: Quantifying uncertainty in range map models", *Ecology Letters*, vol. 9, pp. 1068–1079, 2006.

[JOH 11] JOHNSON K.G., BROOKS S.J., FENBERG P.B. *et al.*, "Climate change and biosphere response: unlocking the collections vault", *BioScience*, vol. 61, pp. 147–153, 2011.

[KAR 11] KARELL P., AHOLA K., KARSTINEN T. *et al.*, "Climate change drives microevolution in a wild bird", *Nature Communications*, vol. 2, p. 208, 2011.

[KAU 08] KAUSERUD H., STIGE L.C., VIK J.O. *et al.*, "Mushroom fruiting and climate change", *Proceedings of the National Academy of Sciences*, vol. 105, pp. 3811–3814, 2008.

[KRE 14] KRELL F.-T., WHEELER Q.D., "Specimen collection: Plan for the future", *Science*, vol. 344, pp. 815–816, 2014.

[LAN 96] LANE M.A., "Roles of natural history collections", *Annals of the Missouri Botanical Garden*, vol. 83, p. 536, 1996.

[LEA 10] LEACHE A.D., FUJITA M.K., "Bayesian species delimitation in West African forest geckos (Hemidactylus fasciatus)", *Proceedings of the Royal Society B: Biological Sciences*, vol. 277, pp. 3071–3077, 2010.

[LEM 81] LEMIEUX L., "The functions of museums", *Gazette*, vol. 14, pp. 4–9, 1981.

[LIS 11] LISTER A.M., CLIMATE CHANGE RESEARCH GROUP, "Natural history collections as sources of long-term datasets", *Trends in Ecology & Evolution*, vol. 26, pp. 153–154, 2011.

[LOM 04] LOMOLINO M.V., *Frontiers of Biogeography: new directions in the geography of nature*, Sinauer Associates Inc., 2004.

[MAD 14] MADDISON D.R., COOPER K.W., "Species delimitation in the ground beetle subgenus *Liocosmius* (Coleoptera: Carabidae: *Bembidion*), including standard and next-generation sequencing of museum specimens: Species of *Bembidion (Liocosmius)*", *Zoological Journal of the Linnean Society*, vol. 172, pp. 741–770, 2014.

[MEI 09] MEIRI S., GUY D., DAYAN T. *et al.*, "Global change and carnivore body size: data are stasis", *Global Ecology and Biogeography*, vol. 18, pp. 240–247, 2009.

[MIL 03] MILLER C.R., WAITS L.P., "The history of effective population size and genetic diversity in the Yellowstone grizzly (*Ursus arctos*): Implications for conservation", *Proceedings of the National Academy of Sciences*, vol. 100, pp. 4334–4339, 2003.

[PAR 04] PARHAM J.F., STUART B.L., BOUR R. *et al.*, "Evolutionary distinctiveness of the extinct Yunnan box turtle (*Cuora yunnanensis*) revealed by DNA from an old museum specimen", *Proceedings of the Royal Society B: Biological Sciences*, vol. 271, pp. 391–394, 2004.

[POL 13] POLGAR C.A., PRIMACK R.B., WILLIAMS E.H. *et al.*, "Climate effects on the flight period of Lycaenid butterflies in Massachusetts", *Biological Conservation*, vol. 160, pp. 25–31, 2013.

[PRI 15] PRICE B.W., HENRY C.S., HALL A.C. *et al.*, "Singing from the grave: DNA from a 180 year old type specimen confirms the identity of *Chrysoperla carnea* (Stephens)", *PLoS ONE*, vol. 10, p. e0121127, 2015.

[PUI 12] PUILLANDRE N., MODICA M.V., ZHANG Y. *et al.*, "Large-scale species delimitation method for hyperdiverse groups", *Molecular Ecology*, vol. 21, pp. 2671–2691, 2012.

[PYK 10] PYKE G.H., EHRLICH P.R., "Biological collections and ecological/ environmental research: a review, some observations and a look to the future", *Biological Reviews*, vol. 85, pp. 247-266, 2010.

[REN 16] RENNER S.S., ROCKINGER A., "Is plant collecting in Germany coming to an end?", *Willdenowia*, vol. 46, pp. 93–97, 2016.

[RUS 08] RUSSELLO M.A., AVERY M.L., WRIGHT T.F., "Genetic evidence links invasive monk parakeet populations in the United States to the international pet trade", *BMC Evolutionary Biology*, vol. 8, p. 217, 2008.

[SCH 15] SCHILTHUIZEN M., VAIRAPPAN C.S., SLADE E.M. *et al.*, "Specimens as primary data: museums and "open science"", *Trends in Ecology & Evolution*, vol. 30, pp. 237–238, 2015.

[SEB 10] SEBASTIAN P., SCHAEFER H., RENNER S.S. *et al.*, "Darwin's Galapagos gourd: providing new insights 175 years after his visit", *Journal of Biogeography*, vol. 37, pp. 975–978, 2010.

[SHA 02] SHAPIRO B., "Flight of the Dodo", *Science*, vol. 295, pp. 1683–1683, 2002.

[SMI 08] SMITH M.A., RODRIGUEZ J.J., WHITFIELD J.B. *et al.*, "Extreme diversity of tropical parasitoid wasps exposed by iterative integration of natural history, DNA barcoding, morphology, and collections", *Proceedings of the National Academy of Sciences*, vol. 105, pp. 12359–12364, 2008.

[SUA 05] SUAREZ A.V., HOLWAY D.A., WARD P.S., "The role of opportunity in the unintentional introduction of nonnative ants", *Proceedings of the National Academy of Sciences of the United States of America*, vol. 102, pp. 17032–17035, 2005.

[THO 09] THOMSEN P.F., ELIAS S., GILBERT M.T.P. *et al.*, "Non-destructive sampling of ancient insect DNA", *PLoS ONE*, vol. 4, p. e5048, 2009.

[UGE 11] UGELVIG L.V., NIELSEN P.S., BOOMSMA J.J. *et al.*, "Reconstructing eight decades of genetic variation in an isolated Danish population of the large blue butterfly *Maculinea arion*", *BMC Evolutionary Biology*, vol. 11, p. 201, 2011.

[WAL 00] WALLANDER E., ALBERT V.A., "Phylogeny and classification of Oleaceae based on rps16 and trnL-F sequence data", *American Journal of Botany*, vol. 87, pp. 1827–1841, 2000.

[WAN 07] WANDELER P., HOECK P.E., KELLER L.F., "Back to the future: museum specimens in population genetics", *Trends in Ecology & Evolution*, vol. 22, pp. 634–642, 2007.

[WAR 07] WARD D.F., "Modelling the potential geographic distribution of invasive ant species in New Zealand", *Biological Invasions*, vol. 9, pp. 723–735, 2007.

[YEA 16] YEATES D.K., ZWICK A., MIKHEYEV A.S., "Museums are biobanks: unlocking the genetic potential of the three billion specimens in the world's biological collections", *Current Opinion in Insect Science*, vol. 18, pp. 83–88, 2016.

Mice and Men: an Evolutionary History of Lassa Fever

Lassa fever is a hemorrhagic viral disease caused by the Lassa virus (LASV), a member of the Arenaviridae family from the Arenavirus genus, and is transmitted to humans by some rodents. Although the Lassa fever was described for the first time in the 1950s, the virus responsible was only isolated in 1969 during an epidemic in Lassa village in Nigeria [FRA 70]. Lassa fever is endemic in West Africa and outbreaks occur in two distinct geographical areas, the Mano river region (SL, L and G) and Nigeria [KEA 77, MCO 87, TOM 87]. However, serological studies suggest that the LASV has recently emerged in various African countries (Senegal 1988; Mali 2009; Burkina Faso 2010; Ivory Coast 2000, 2013; Benin 1977, 2014; Cameroon 2011; Ghana 2011; Democratic Republic of Congo 2011; Uganda 2017) [OMS 17]. LASV is also the hemorrhagic fever most frequently exported to countries of the Global North, with more than thirty cases documented since 1969: in England (1971, 1972, 1975, 1976, 1981, 1982, 1984, 1985, 2000, 2003), the Netherlands (1980, 2000), Germany (1974, 2000, 2006, 2016), Sweden (2016), Israel (1987), the United States (1969, 1975, 1976, 1989, 2004, 2015), Canada (1989) and Japan (1987) [OMS 17].

Mortality rates due to LASV vary depending on the strain of virus studied, the diagnostic methods, the level of healthcare in hospitals, the sources and modes of infection, the infecting dose, the high degree of variability among infected people and the level of virulence required for the disease to be dealt with. However, it is estimated that LASV affects 300,000

Chapter written by Aude LALIS and Thierry WIRTH.

to 500,000 people a year in West Africa with a mortality rate that can reach 15% [OMS 17]. Human seroprevalence in endemic zones can reach more than 50% [BIR 01]. With yellow fever, dengue fever and the Ebola virus, LASV is one of the most highly dangerous and contagious fevers with the potential to spread. LASV therefore constitutes a considerable health problem in West Africa. As recently as August 2015, an outbreak in Nigeria lead to more than 2,500 presumed cases and around a hundred deaths [OMS 17]. Between January and May 2016, LASV emerged in Benin with 318 monitored patients, 54 confirmed cases and 28 deaths. Although the end of the outbreak was declared on the 23 May 2016, two deaths occurred in February 2017. Furthermore, 2 alleged cases were announced in North Togo and one case was reported in Burkina Faso in February 2017.

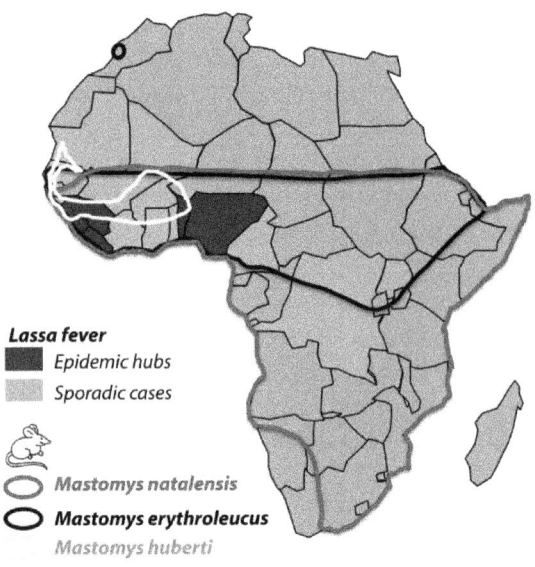

Figure 11.1. *LASV and Mastomys rodent host distribution map. For a color version of this figure, see www.iste.co.uk/grandcolas/biodiversity.zip*

11.1. Symptoms and methods of prevention

In general, human contamination occurs by a bite and by contact with rodent secreta and excreta; sometimes by inhalation of small infected particles or by ingestion (in certain regions, rodents are eaten). Inter-human transmission occurs upon direct contact with the blood, urine, feces or other biological liquids of an infected person. Sexual transmission has also been seen [OMS 17]. Lassa fever shows great variation in terms of clinical signs, symptoms and level of virulence [FRA 70, WHI 72, MER 73, MON 74, MON 84, MCO 87]. Lassa virus infection has been observed with no age or sex differences. The people at greatest risk are those who live in suburban and rural areas (neglected sanitation and overpopulated areas favor the presence of rodents). Although Lassa fever is considered as a viral hemorrhagic fever, the bleeding manifestations are relatively rare. In around 80% of cases, human infection remains asymptomatic; in other cases, several organs are severely affected: the liver, spleen and kidneys. The incubation period varies from 6 to 21 days. The beginning of clinical manifestations is usually progressive, with fever, a general sensation of weakness and nausea. After a few days, the infected person may start to suffer from headaches, throat irritation, thoracic pains, vomiting, diarrhea, coughing as well as abdominal pains [OMS 17]. Recovery is completed in most cases. Loss of hearing is the major side-effect of the disease and may be irreversible in 29% of infected subjects [CUM 91]. It is also considered that LASV is the most common cause of deafness in several countries in West Africa [LIA 92]. In the most serious cases of infection, a facial edema, liquid in the lung cavity, hemorrhages in the oral and nasal cavities and the digestive system, and hypotension can appear [OMS 17]. At a late stage, a rapid deterioration has been observed (less than 14 days) progressing towards a shock, convulsions, neurological manifestations sometimes accompanied by encephalopathies with respiratory distress, coma, convulsions and death [MCO 87]. The clinical diagnosis of LASV is difficult; the first symptoms can often be similar to those of other illnesses present in the tropics such as malaria, typhoid fever, leptospirosis and flu. The best method to diagnose Lassa fever consists of serological tests combining the detection of antibodies and antigens (Elisa technique). The Lassa antigen is in general present in the blood a few days after first infection and disappears upon the arrival of antibodies, which themselves appear a few days after infection and reach a maximum after 10–12 days of infection. Although their lifespan has not yet been extensively studied, these antibodies are present for years [MCO 87]. There is currently no vaccine available against LASV in humans, and the sole treatment, relying on the drug ribavirin, is only effective if administered early in infection, i.e. within the first 6 days after the onset of clinical symptoms. In fact, as already explained, the symptoms found at the start of the illness being similar to those observed in other very frequent pathologies in these regions (malaria, dysentery), the presence of LASV is often only thought of several days after the first appearance of symptoms.

Ribavirin, in the rare cases where it is available and accessible, is therefore usually administered too late to be effective. Some studies are currently being undertaken in order to develop a LASV vaccine. The development of a Lassa fever vaccine has been proposed [FIS 01], but its effectiveness was very quickly shown to be limited by the existence of numerous viral strains of varying virulence and by the economic and social barriers that slow down development and distribution of vaccine candidates in West Africa [RIC 03].

Currently, the main methods to design a LASV control are the isolation of infected persons, the disinfection and surveillance of people who have had close contact with a patient, and rodent management. For example, people in endemic zones must ensure that rodents do not get into the household and enter into proximity with food stocks. The World Health Organization (WHO) has developed a publicity campaign on 'Knowing Lassa fever and protecting yourself' [OMS 16] in several African countries by releasing huge awareness campaigns on basic rules of hygiene, in order to inform urban and rural populations about the LASV risks. The government of Benin has recently developped a cartoon about LASV and organized some seminars about the LASV disease in the schools of infected areas.

Furthermore, the search for pathogens (viral screening) in rodents must be systematically carried out in order to make possible an epidemiological surveillance of host rodents. Until recently, only four arenaviruses were known in Africa (Ippy, Mobala, Mopeia and Lassa) even though the genus comprises twenty or so species distributed for the most part in America. However, since a few years ago, several arenaviruses have been identified in Africa: the Kodoko virus in 2007 in dwarf mice (*Mus Nannomys minutoides*) in Guinea [LEC 07]; the Mopeia virus in Tanzania in 2009 in Natal multimammate mice (*Mastomys natalensis*) [GUN 09]; the Lujo arenavirus identified in South Africa in 2009, the reservoir of which is still unknown today [BRI 09]; and finally, in Ivory Coast, the arenaviruses Gbagroube and Menekre related respectively to LASV and to the Ippy/Mobala/Mopeia virus complex in 2011 in bush mice (*Hylomyscus sp.*) and the dwarf mouse (*Mus Nannomys setulosus* [COU 11]). The recent discovery over the last few years of new arenaviruses shows that research is often rewarded with success when led by scientists in possession of the appropriate tools and technology. This line of thought can only be considered in terms of international teamwork operating on a long-term basis. Without underestimating the harmful effect of rodents in developed countries, it is nonetheless very clear that the need to control these species is more urgent in developing countries, which are heavily affected by periods of climatic variation (drought, floods, etc.). Furthermore, public health conditions in these countries are particularly worrying and the impact of rodents is a determining factor. The development of strategies of rodent control will depend on incoming knowledge about the dynamic of host species, aid projects brought to local populations and the support given to simple preventative sanitary measures. The study of rodent-borne viruses and their interactions with their hosts is essential, not only because of the seriousness of the diseases they cause, but also the potential outbreaks they represent.

Figure 11.2. *School intervention program on LASV in an elementary school in Tchaourou (Benin) (© UNICEF Benin) and example of an LASV awareness campaign cartoon (© Government of Benin). For a color version of this figure, see www.iste.co.uk/grandcolas/biodiversity.zip*

11.2. Towards a better understanding of the viral reservoir: phylogenetic systematic gene flow and phenotypic variation

The infection reservoir is thus represented by a peridomestic rodent from the *Mastomys* genus, which belongs to the Muridae family (rats, mice, hamsters, voles, lemmings and gerbils). The vernacular name for *Mastomys* is "Natal multimammate rat". The common name describes the high number of mammary glands, usually 8 to 12 pairs, but there may be as many as 18 pairs. This number is much larger than what is normally found in other murid species (in general, 3 to 5 pairs) [PET 77] and allows higher fertility for the *Mastomys* genus. The *Mastomys* genus is generally commensal,

African endemic, naturally coming from the savanna and spread throughout the whole of Sub-Saharan Africa (Figure 11.1). The genus is capable of occupying different habitats: forests, crop fields or uncultivated land, village or small town habitations [DUP 88]. *Mastomys* is considered as a pest, causing a lot of damage to crops [LEI 97]. It is also characterized by frequent spreading and rapid recolonization of habitats [LEI 96, LEI 97, MWA 02, GRA 05]. Other than the damages in cultivated land and crops, their interactions with human populations can also have significant consequences on public health. In fact, as well as being the reservoir and host of the LASV, some species of *Mastomys* are natural reservoirs and/or vectors of parasites and zoonoses such as different salmonella serotypes (gastroenteritis, hepatitis, meningitis, arthritis, etc.), spotted fever and Q fever (pneumonia) [CON 81], as well as Leptospirosis [HOL 06].

The phylogenetic systematic and evolutionary history of the *Mastomys* genus have been confused for a long time to the point of combining different methodological approaches because the *Mastomys* genus includes twin species that are morphologically similar but with different genetic structures, isolated from one another in term of reproduction [MAY 63]. The species of this genus can only be determined unambiguously after resorting to molecular barcode and chromosomal data because each species has a specific karyotype [DUP 90, GRA 97, LAV 98, VOL 98, Vol 01, DOB 02]. Out of the eight species in the genus, three are found in West Africa: *M. natalensis*, *M. erythroleucus* and *M. huberti* [MUS 05]. The specific determination of the LASV reservoir has never been very clear between the three main species of the genus who have sympatric areas (overlapping distribution): *Mastomys natalensis* was first identified as a LASV vector, then *M. erythroleucus* and *M. huberti* species were suspected [MON 75, MCO 87]. In 2006, *M. natalensis* was identified as a natural host of LASV in Guinea [LEC 06]. Very recently, *M. erythroleucus* (as well as the mouse *Hylomyscus pamfi*) was recognized as an alternative host to *M. natalensis* in Guinea [EFC 16] and in Nigeria [OLA 16]. This demonstrates that the virus is genetically and ecologically complex, possibly with several reservoirs. Furthermore, these rodent species have different histories and lifestyles and occupy a range of ecological niches making the potential outbreak and spread of LASV more likely.

LASV has not been discovered in Central, East or South Africa, even though *M. natalensis*, the most widespread species of the genus, is observed. This situation could be due to the absence of the virus in rodent populations outside of endemic zones. However, these data could equally underline that the distribution we currently recognize as accurate is incomplete as there is never enough systematic virological monitoring or surveys on rodent populations. Thus, the cartography of LASV in Africa and the *M. natalensis* populations has been modeled and risk of LASV circulation spreading outside the recognized endemic area has been evaluated [MYL 15]. The results predicted that 37.7 million people in 14 countries, mainly in West Africa, live in areas where all conditions for the transmission of LASV between rodents and humans to take place are present (presence of *M. natalensis*, vegetation, temperature, altitude, evapotranspiration). Five of these countries (Niger, Senegal, Guinea-Bissau, Cameroon and Togo) where the human populations are considered at risk had not yet shown cases of LASV at the time of the study in 2015. Since, Togo has had two deaths [OMS 17]. Climate change, deforestation and the intensification of commercial exchange in Africa are factors in the emergence of this infectious disease.

Mastomys natalensis is a commensal species and therefore easily found in housing and near stocks of food and water [MCO 87, DUP 88, CHI 93]. Although *Mastomys* are nocturnal, closed houses or houses without electricity in African villages make it possible for rodents to be active throughout the day, thereby increasing the frequency of contact between humans and rodents. Like all arenaviruses, LASV persists in nature in the form of a rodent-rodent cycle and persistent infection in *Mastomys* [CHI 93]. Vertical transmission (from mother directly to embryos) of the virus results in an asymptomatic and persistent infection in the next generation. Infected animals are viremic (presence of virus in the blood) and excrete the virus in urine and faeces. On the contrary, horizontal transmission (contact with an infected individual) between adult animals causes a transient and immunizing infection. The infection's effect on the host rodent appears negligible. Human contact with rodent excretions (touching the excretions or eating food contaminated with excretions from rats) is the predominant mode of infection [MON 75]. In endemic zones, infection risks are linked to the frequency of contact with rodents. Examples of high-risk situations causing a high density of rodents are human overpopulation (economical and political reasons), poor storage of food and hunting (bush meat) of rodents as a source of food [FRA 74, TER 96]. Although LASV is seen all year long in endemic zones, infection is more frequent during the drought season due to increased contact between people and rodents that become concentrated in houses close to stocks of food and water [FIC 07].

Despite the seriousness of the threat of LASV, the rapid risk assessment of interhuman transmission and the potentially significant emerging risk by the rodent host distribution, the knowledge about the reservoir Mastomys (dipersion, population dynamic...) remain very limited. However, thanks to the European Union's IncovDev project (N° ICA4-2002-10050) led by EVEE team from the ISYEB MNHN and to its eco-epidemiological approach (prevalence of LASV: environmental factors, rodent communities and ecological characteristics such as specialization in terms of habitat and commensalism), *M. natalensis* populations have been studied on a country-wide scale. Therefore, for the first time, thanks to collaboration between systems analysts, epidemiologists, genetic scientists and French and Guinean doctors, more than 3000 rodents have been successfully captured [LEC 06] and monitored (2002–2008) [FIC 05, FIC 07, FIC 08, FIC 09]. With some medical consultations carried out in villages, we have been also able to gather precise data on the distribution of LASV in rodent and human populations as well as to unravel *M. natalensis* phenotypic and genetic variability [LAL 08, LAL 12, LAL 15] in order to better understand its capacity to adapt and spread.

Figure 11.3. *Top left: IncoDev Project:* Mastomys natalensis © Aude Lalis; top right: Sherman Trap used to catch rodents © Aude Lalis; bottom left: brush autopsy camp: dissection and samples © Stéphane Bégoin; bottom right: medical consultation © Stéphane Bégoin. For a color version of this figure, see www.iste.co.uk/grandcolas/ biodiversity.zip

The results of the research project revealed that distribution of the *M. natalensis* species in Guinea and its relative abundance corresponded to the geographical zones where the highest human LASV seroprevalence was observed (southern borders with Sierra Leone and Liberia) [LUK 93, DEM 01]. Infected *M. natalensis* are from two regions at risk in Guinea: the sub-prefecture Faranah (Upper Guinea: localities of Tanganya, Bantou and Gbetaya) and the Dengeudou region (forested Guinea), not far from the Kenema region, home to hemorrhagic fever in Sierra Leone where epidemics are frequent.

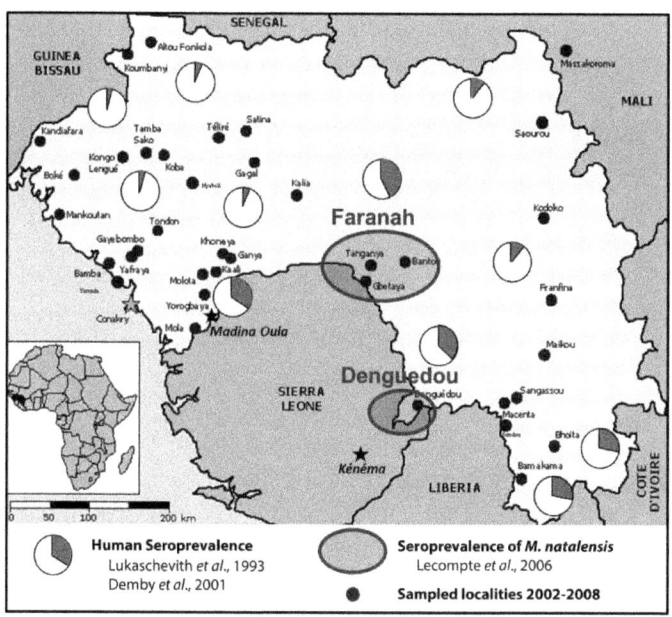

Figure 11.4. *Mapping the human and rodent LASV seroprevalence in Guinea [LUK 93, DEM 01, LEC 06]* © Aude Lalis

It is interesting to note that the high human prevalence rate measured in 2001 in Maritime Guinea is observed in an ancient refugee camp (Madina Oula) where the only LASV epidemic known in Guinea occured in 1982 [BOI 87]. A recent study reveals the capture of six specimens of LASV positive *M. erythroleucus* in Madina Oula [OLA 16], which became an alternative host after an adaptation of the LASV [OLA 16]. The main, highly commensal host species, *M. natalensis*, is found predominantly in the forest ecosystems in the south-east of the country, but the species' additional presence in the north-east and south-west regions represents a risk for the spread of LASV, especially if degradation of the forest environment caused by human exploitation persists. The intensity of deforestation could provide *M. natalensis* with new spaces in which to evolve.

Although there is a strong interest for rodent population control, especially in order to understand processes of disease colonization and transmission to humans, genetic variability and gene flow in the *Mastomys* genus are not well known. There have been very few studies published on the spatial genetic structuration in *Mastomys* [BRO 07]. However, within the framework of this research carried out in Guinea, we have been able to show that the geographical and commensal specializations in *M. natalensis* explain the high level of genetic differentiation measured between populations, the absence of strong gene flow and the low dispersion level over a large territory [LAL 12]. Thus, dispersion of *M. natalensis* is reduced in less inhabited zones. In fact, the species' dispersion seems limited to localities close to one another and is characterized by low levels of migration. Dispersion by a human transport vector is therefore probably very limited as no migrant *M. natalensis* has been observed over long distances, for example, along main communication routes. *M. natalensis* is strictly confined to human habitations (and nearby cultivated gardens), just as in Mali [GRA 97, 05] and Senegal [DUP 90, BRO 07]. The commensal habitat offers some favorable situations in terms of constant availability of food resources and stable environmental conditions, which no doubt engenders sedentary behavior and reproductive activity throughout the year. Survival problems for this omnivorous-granivorous species do not stem from the quality of food itself, but its availability, that is to say the energy that must be spent in search of food [HUB 81]. The permanent availability of food in human habitations modifies individuals' metabolism and is no doubt the reason for which an increase in food resources is followed by an increase in the length of survival of these animals and a decrease in the size of their home range, the consequence of which is a local increase in population density of the animal [HUB 81]. It has been observed that population dynamics and reproduction differed between two geographically close villages [FIC 07]. This observation is linked to the level of poverty and to the difference in food stock quantity found in both villages. The quality and quantity of available food therefore condition the development of rodent populations and their distribution. As such, social structure directly influences the genetic structure of populations. In *M. natalensis* in Senegal, the existence of familial groups, consisting of a male adult, several female adults and a varying number of juveniles and sub-adults, has been highlighted [GRA 86]. These groups define a reproduction unit called a deme and the distribution is conditioned by the human habitat. Within one site, *M. natalensis* has only rarely been caught outside of human habitations [LAL 08]. This species probably has difficulty moving about in open environments, which limits its spread even on a small local scale. As a consequence, although highly commensal, *M. natalensis* does not seem to spread by the vector of human transport and is characterized by rare events of dispersion over long and short distances [LAL 12]. Thus, the spread of the species is probably limited by environmental physical properties (distance to cover, forest environments not favorable to the species, rarifaction of villages in the north of the country) and perhaps by competition and predation pressures of more significance in open environments [BRO 07]. Commensal specialization and migration rates among host populations of *M. natalensis* have allowed us to estimate the zones at risk to LASV outbreaks in Guinea and to reinforce the knowledge and understanding necessary to establishing effective strategies to prevent this viral disease.

While the association between the *Mastomys* genus and LASV has been described for more than 30 years ago, few studies have described the host/LASV relationship. Many studies on the effects of arenaviral infection (lymphocytic choriomeningitis virus -LCMV, Pichinde virus) on the level of brain development in rats have made it possible to obtain pertinent models to understand the host/virus interaction [BAL 93, WRI 95, SHA 02, RAM 03]. These studies have brought some new knowledges of the pathological processes of chronic infections, specially in the size and development of certain parts of the brain (cerebellum, choroid plexus, hippocampus and olfactory bulbs) [BON 02]. Hence, it seemed interesting to us to study the influence of the viral infection LASV on the phenotypic variability of its main host, *M. natalensis*, thanks to a geometric morphometry study (study and analysis of the form of a structure/organ) [LAL 15]. This study carried out on cranial structure indicated that there was a link between the presence of LASV and differences in size and cranial conformation in *M. natalensis* [LAL 15]. In fact, these results revealed a morphological differentiation between two groups of *Mastomys* (infected versus uninfected by LASV). This differentiation could reflect that LASV positive specimens belong to one genetic lineage and/or a phenomenon of development instability (inconsistent production of the phenotype) arising following vertical transmission of LASV [LAL 15].

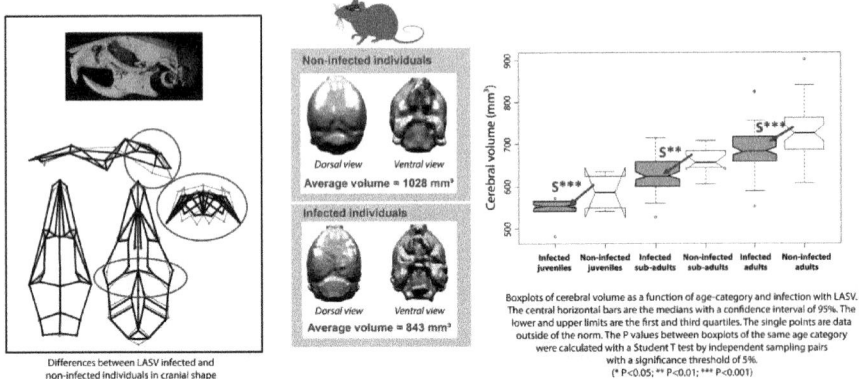

Figure 11.5. *Measure of cranial phenotypic plasticity in* M. natalensis *in a high prevalence area (Faranah region) by geometric morphometry © Aude Lalis. For a color version of this figure, see www.iste.co.uk/grandcolas/biodiversity.zip*

COMMENT ON FIGURE 11.5.– Differences between LASV infected (BLACK) and non infected individuals (GREY) in cranial shape; non-infected individuals; dorsal view; ventral view; average volume; infected individuals; dorsal view; ventral view; average volume; differences in cerebral volume between groups of infected and non-infected individuals; cerebral volume; infected juveniles; juveniles; infected sub-adults; sub-adults; infected adults; adults; boxplots of cerebral volume as a function of age-category and infection with LASV. The central horizontal bars are the medians with a confidence interval of 95%. The lower and upper limits are the first and third quartiles. The single points are data outside of the norm. The P values between boxplots of the same age category were calculated with a Student T test by independent sampling pairs with a significance level of 5%.

In order to verify the hypothesis of a marked cerebral hypoplasia (insufficient growth) in LASV infected individuals and thus confirm the phenomenon of developmental instability observed in M. natalensis, we estimated brain case volume. We have shown that individuals infected with LASV showed a significant reduction (nearly 20%) of the volume occupied by cerebral mass [LAL 15]. These results have thus revealed a developmental instability probably occurring during the gestation period when vertical transmission of LASV occurs. In fact, the female transmits the virus to embryos, "healthy" carriers nonetheless chronically infected with LASV for life (with side-effects on growth and behavior). Furthermore, infected individuals showed a significant reduction in body size, isometric size and the size of reproductive organs [LAL 15]. We think these morphological changes are the consequence of a reduction in the activity of a growth hormone causing a decrease in the selective value of infected individuals. We hypothesize that growth impairment may result in a selective disadvantage for LASV-infected *M. natalensis*, leading to a preferably commensal lifestyle in areas where the LAVS is endemic and, thereby, increasing the risk of LASV transmission to humans. Our results highlight an imbalanced host/virus relationship with an imperfect adaptation between the host species *M. natalensis* and LASV [LAL 15].

The mechanisms acting on behavioral changes after infection in vertebrate hosts are varied. Several pathogens, including viruses such as the Borna disease virus (BDV) and Rabies virus, can penetrate the central nervous system and cause modifications in the cerebral regions known for their involvement in the expression of emotional, cognitive and social behavior (hippocampus, cerebellum) [JOH 70, ARN 90, Sol 95, HAT 97, HOL 01, MOO 02]. During host–virus coevolution, host populations developed adaptations to avoid infection and the pathogenic viruses developed counter-adaptations to bypass the defense mechanisms of host populations. In many cases, these counter-adaptations involved a direct manipulation of the host's behavior to increase contact between infected individuals and those most sensitive to the pathogen [MOO 02]. Studies have shown changes in social behavior in infected host populations, notably in several rodents (*Peromyscus boylii, Peromyscus maniculatus, Reithrodontomys magalotis, Clethrionomys glareolus, Stigmodon hispidus, Rattus norvegicus*) [COM 01, MOO 02, KLE 02, 03]. Thus, different types of pathogenic viruses would be able to manipulate mechanisms that directly control the expression of social behaviors, such as aggressiveness that would make the transmission of these viruses easier [MOO 02, KLE 03]. It has also been shown that neonatal BDV infection was associated with anomalies in the social behavior of young and adult rats. Alterations in social activity due to changes in the emotional and cognitive domains of the brain are linked to damage during cerebral development, particularly the degeneration of the hippocampus and of certain cells in the cerebellum. Infected rats then spend less time in active social interactions and more time pursuing sexual partners. The authors have therefore highlighted that infection by BDV alters the normal pattern of social interaction among rats. They have revealed the important role the host/virus BDV model could play in the study of cerebral neuro-lesions and in the pathology of developmental anomalies such as autism [LAN 07]. In the case of the *M. natalensis*/LASV model, it would be interesting to carry out more in-depth research on the neurological consequences of infection, on the level of the central nervous system but above all on the level of the brain, in order to more precisely determine which behavioral changes favor transmission of the virus. Here, we are perhaps dealing with a context of vertical transmission of LASV which would cause a reduction in the activity of a growth hormone reducing the selective value of infected individuals, and would cause neuro-cerebral anomalies. Furthermore, behavior relating to exploration of the immediate environment in infected individuals could be altered as suspected in the case of viral infection of rats in laboratories by the arenavirus LCMV (lymphocytic choriomeningitis virus) [BON 02]. The transmission of LASV to humans would then be favored by a more pronounced tendency towards a commensal lifestyle in infected individuals and by the constitution of infected demes according to a model: one or a small number of huts would accommodate a deme of infected *M. natalensis* [LAL 12]. The results of the study of genetic diversity in infected individuals has allowed us to verify that individuals carrying the virus showed significantly higher rates of consanguinity compared with non-infected individuals but did not seem genetically isolated within populations [LAL 15]. Additionally, horizontal transmission of the virus could be favored by an increase in aggressive behavior in infected individuals [KLE 04]. The most aggressive individuals would then be dominant and have greater access to resources. If dominance is linked to a higher reproductive status [LAN 07], then dominant males would engender more descendants than submissive males. Yet this hypothesis seems incompatible with the detected decrease in size of reproductive organs [LAL 15]. In order to verify these hypotheses, more comprehensive research must be carried out on this complex system of host/virus coevolution that still remains under-explored. The results of the IncoDev project may be considered as an exploratory study making it possible to reveal basic factors that are informative and essential for a better understanding of the evolutionary biology of *M. natalensis* populations in the context of viral infection. The arenaviruses, like LASV, lead to neurological complications in infected animals but also in people [RAM 03]. *Mastomys natalensis* could therefore make a model for the examination of patho-morphological mechanisms of brain development upon viral infection of a natural host, with the aim of yielding on a long-term basis more decisive understanding in the domain of human virology.

11.3. Lassa virus, molecular evolution and dating

The genome of the virus responsible for LASV is of modest size (10.7kb), and consists of a single-stranded RNA molecule divided into two segments, respectively, called S and L (short and long). The first segment codes for nucleoprotein and for surface glycoproteins precursor (GPC), and the second codes for protein Z and an RNA polymerase (L). Nearly 10% of rodents captured in Guinea have proved to be positive based on RT-PCR analysis (technique in molecular biology that allows the amplification of viral RNA) [LEC 06]. These data combined with published sequences since 1969 in the total area of distribution have allowed us to carry out a phylo-geographic analysis of LASV. Phylogenetic analysis confirms the presence of two main evolutionary lineages overlapping the two main endemic basins. On the one hand, Nigeria to the east and, on the other hand, the "Mano River" region to the west. The root of the phylogenetic tree based on the initial strain collected in 1969 strongly suggests that this hemorrhagic fever is of Nigerian origin, which information is backed up by a higher nucleotide diversity found in Nigeria compared with that of the "Mano River" region.

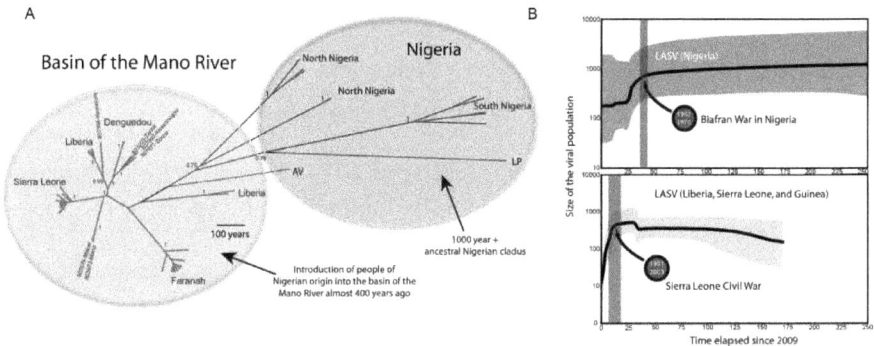

Figure 11.6. *Diversity and genetic structure of LASV. a) Phylogenetic tree of Lassa virus based on nucleotide sequences of the nucleoprotein (NP); b) demographic fluctuations of the viral population over time © Thierry Wirth. For a color version of this figure, see www.iste.co.uk/grandcolas/biodiversity.zip*

The majority of RNA viruses, such as Ebola virus or HIV cumulate mutations in a linear fashion over time. As a consequence, phylogenetic branches of the viruses reconstructed from gene sequences get longer over time. A sample corresponding to a strain from 1970 will therefore have a shorter branch than a strain from the 1990s, which itself will be shorter than the branch of a contemporary strain. Thus, when we draw a correlation between branch length and the year of a strain, the slope corresponds to the rate of mutation. The Bayesian algorithm BEAST, based on these properties, makes it possible to infer the mutation rate of a given gene, to date the age of a line of descent, but also to trace the demographic changes of the population over time. The deduced rate of molecular evolution of the nucleoprotein (NP) is 1.6×10^{-3} substitution/nucleotide/year, an estimation situated in the middle range of negative RNA viruses. The common ancestor of the Nigerian line of descent is estimated to be around 900 years, while the common ancestor of strains from West Africa only dates back to around 200 years, which confirms the late and secondary transfer of strains that came from Nigeria towards West Africa [LAL 12]. However, we must bear in mind that the estimation of the dating of common ancestors is dependent on good sampling, as the addition of missing basal strains can strongly affect the estimation. It is therefore no surprise that a recent study on nearly 200 strains from Sierra Leone and Nigeria has produced results that are slightly different from ours. Andersen and his colleagues report a common ancestor of LASV dating back 1060 years and a secondary arrival in the Ivory Coast roughly 300 years ago [AND 15]. The older the common ancestor, the more biased the estimation is, which means in the case of LASV that the dating of its arrival in West Africa is fairly reliable, while that of the common ancestor of all strains is underestimated, several thousand years being more than plausible.

When studying more precisely the demographic changes in LASV's two main evolutionary lines of descent, one of the most striking facts is the contrasting pattern of these two lineages. The viral population in Nigeria has remained relatively constant during the last 250 years and underwent a slight decrease roughly 50 years ago, followed by stabilization. On the contrary, strains from the "Mano River" basin went through a long period of stasis before a sharp drop (by a factor of 500) of the effective size of the viral population around 20 years ago [LAL 12].

11.4. Role of armed conflicts and refugee movements in the spread of Lassa fever in West Africa

At first sight, the demogenetic inferences about the viruses from the "Mano River" basin are somewhat surprising since they suggest a sharp demographic decrease in the pathogen during the outbreak periods. To recapitulate, Lassa fever is endemic in these regions and numerous epidemic episodes with high mortality rates occurred between 1971 and 2000. The civil war in Sierra Leone from 1991 to 2003 came with an intense and constant flux of migrants (more than 500,000 people forced to move) towards Guinea, which was more politically and socially stable, thus making of this country a leader in Africa in terms of its quota of refugees. The fall in size of the viral population can be explained by two man-made phenomena that are independent yet potentially coordinated: on the one hand, by deforestation and ecological disturbances caused by population movements and the installation of refugee camps, the effects of which are clearly illustrated by satellite images showing the region of the Denguedou camp before and after its construction (https://earthobservatory.nasa.gov/IOTD//view.php?id=6450) and, on the other hand, by the consumption of rodents as a ready source of protein. This interpretation is reinforced by mathematical analysis of the demographic dynamics of *Mastomys natalensis*, rodents in the regions concerned showing a huge demographic drop during the period in question [LAL 12]. This rodent population that lives close to camps is also atypical in the sense that it is genetically different from other *M. natalensis* populations sampled in Guinea [LAL 12]. A credible explanation is that refugees from Sierra Leone or Liberia accidentally brought infected rodents with them, notably by transporting sacks of grain, thus spreading the disease towards Guinea. The impact of conflicts on the spread of Lassa fever also seems to be a process that has repeated itself, since a similar situation occurred in Nigeria, where the viral population fell during the Biafra war. These signs all converge towards a certain interpretation that, while lacking in ultimate proof, points towards one of the most likely processes responsible.

11.5. Viral epidemiology: towards an integrated approach

The combination of conclusions we have been able to present in our work [LEC 06, LAL 12, LAL 15] was only possible thanks to the use of an integrative approach. In fact, only by combining the analysis of rodent genotypes and those of their viruses, bringing the genetics of whole populations, phylogeography and Bayesian tools together in a unified framework of analysis, have our interpretations been able to converge. Taken individually, our results would have led us to false interpretations. Unfortunately, and too often still, studies dedicated to hemorrhagic viruses focus solely on strains that have come from carriers (sampled during the period in which subjects are carriers of the virus and can transmit hemorrhagic fever) or human patients, thus biasing sampling while obscuring a large part of the evolutionary history of the pathogenic agent in its natural hosts such as rodents. Finally, another lesson to learn is the strong genetic variability and geographic structure of LASV, which dates back several thousand years; a pattern that questions the success of a unique vaccination. According to our results, an approach of diagnosis and vaccination specific to each country and even to specific regions will perform far better.

Emerging infectious diseases such as LASV remain one of the main causes of death in the world. The evolution of lifestyles, the appearance and fast spread of viral strains, the emergence of new pathologies and the outbreak of large pandemics (tuberculosis, Ebola, SRAS, H5N1 bird flu, etc.) show the need to better understand the phenomenon of infectious diseases in its totality in order to: i) answer fundamental research questions about host/virus interactions and the emergence of viral diseases, while remaining within the precise framework of research applied to the prevention of new and rare diseases; ii) develop new therapeutic and diagnostic tools ahead of future public health problems; and iii) to improve surveillance and prevention. The emergence of new infectious diseases is a threat to humankind that must be anticipated. A better understanding of the biological mechanisms that lead to the appearance of new diseases in humans transmitted by host/reservoir animal species is essential if we are to be equipped with the tools necessary to control the viral epidemics to come.

11.6. Bibliography

[AND 15] ANDERSEN K.G., SHAPIRO B.J., MATRANGA C.B., "Clinical sequencing uncovers origins and evolution of lassa virus", *Cell*, vol. 162, no. 4, pp. 738–750, 2015.

[ARN 90] ARNOTT M.A., CASSELLA J.P., HAY J., "Social interactions of mice with congenital Toxoplasma infection", *Annals of Tropical Medicine and Parasitology*, vol. 84, pp. 149–156, 1990.

[BAL 93] BALDRIDGE J.R., PEARCE B.D., PAREKH B.S. *et al.*, "Teratogenic effects of neonatal arenavirus infection on the developing rat cerebellum are abrogated by passive immunotherapy", *Virology*, vol. 197, pp. 669–677, 1993.

[BIR 01] BIRMINGHAM K., KENYON G., "Lassa fever is unheralded problem in West Africa", *Nature Medicine*, vol. 7, p. 878, 2001.

[BOI 87] BOIRO I., LOMONOSSOV N.N., SOTSINSKI V.A. *et al.*, "Eléments de recherches clinico-épidémiologiques et de laboratoire sur les fièvres hémorragiques en Guinée", *Bulletin de la Société de Pathologie Exotique*, vol. 80, pp. 607–612, 1987.

[BON 02] BONTHIUS D.J., MAHONEY J., BUCHMEIER M.J. *et al.*, "Critical role for glial cells in the propagation and spread of lymphocytic choriomeningitis virus in the developing rat brain", *Journal of Virology*, vol. 76, no. 13, pp. 6618–6635, 2002.

[BRI 09] BRIESE T., PAWESKA J.T., MCMULLAN L.K. *et al.*, "Genetic detection and characterization of lujo virus, a new hemorrhagic fever-associated arenavirus from Southern Africa", *PLoS Pathog*, vol. 5, no. 5, p. e1000455, 2009.

[BRO 07] BROUAT C., LOISEAU M., KANE M. *et al.*, "Population genetic structure of two ecologically distinct multimammate rats: the commensal Mastomys natalensis and the wild Mastomys erythroleucus in southeastern Senegal", *Molecular Ecology*, vol. 16, no. 14, pp. 2985–2997, 2007.

[CHI 93] CHILDS J.E., PETERS C.J., "Ecology and epidemiology of arenaviruses and their hosts", in SALVATO M. (eds), *The Arenaviridae*, Plenum Press, New York, 1993.

[COM 01] COMBES C., *Parasitism: The Ecology and Evolution of Intimate Interactions*, University of Chicago Press, Chicago, 2001.

[CON 81] CONWAY G., "Man versus pests", in MAY R., MCLEAN A. (eds). *Theoretical Ecology: Principles and Applications*, Oxford University Press, Oxford, 1981.

[COU 11] COULIBALY-N'GOLO D., ALLALI B., KOUASSI S.K. *et al.*, "Novel arenavirus sequences in Hylomyscus sp. and Mus (Nannomys) setulosus from Côte d'Ivoire: implications for evolution of arenaviruses in Africa", *PLoS One*, vol. 6, no. 6, p. e20893, 2011.

[CUM 91] CUMMINS D., MC CORMICK J.B., BENNET D., "Acute sensorineural deafness in Lassa fever", *Journal of the American Medical Association,* vol. 264, pp. 2093–2096, 1991.

[DEM 01] DEMBY A.H., INAPOGUI A., KARGBO K. *et al.*, "Lassa fever in guinea: II. distribution and prevalence of lassa virus infection in small mammals", *Vector-Borne and Zoonotic Diseases*, vol. 1, pp. 283–296, 2001.

[DUP 88] DUPLANTIER J.M., GRANJON L., "Occupation et utilisation de l'espace par des populations du genre Mastomys au Sénégal : étude à trois niveaux de perception", *Sciences et Techniques de l'Animal de Laboratoire*, vol.13, no. 2, pp. 129–133, 1988.

[DUP 90] DUPLANTIER J.M., GRANJON L., MATHIEU E. *et al.*, "Structures génétiques comparées de trois espèces de rongeurs africains du genre Mastomys au Sénégal", *Genetica*, vol. 81, pp. 179–192, 1990.

[DOB 02] DOBIGNY G., BAYLAC M., DENYS C., "Geometric morphometrics, neural networks and diagnosis of sibling Taterillus (Rodentia, Gerbillinae) species", *Biological Journal of Linnean Society*, vol. 77, pp. 319–327, 2002.

[FIC 05] FICHET-CALVET E., KOULEMOU K., KOIVOGUI L. *et al.*, "Spatial distribution of commensal rodents in regions with high and low Lassa fever prevalence in Guinea", *Belgian Journal of Zoology*, vol. 135, pp. 63–67, 2005.

[FIC 07] FICHET-CALVET E., LECOMPTE E., KOIVOGUI L. *et al.*, "Fluctuations of abundance and Lassa virus prevalence in Mastomys natalensis in Guinea, West Africa", *Vector-Borne and Zoonotic Diseases*, vol. 7, no. 2, pp. 119–128, 2007.

[FIC 08] FICHET-CALVET E., LECOMPTE E., KOIVOGUI L. *et al.*, "Reproductive characteristics of Mastomys natalensis and Lassa virus prevalence in Guinea, West Africa", *Vector-Borne and Zoonotic Diseases*, vol. 8, no. 1, pp. 41–48, 2008.

[FIC 09] FICHET-CALVET E., ROGERS D.J., "Risk Maps of Lassa Fever in West Africa", *PLoS Negl Trop Dis*, vol. 3, no. 3, p. e388, 2009.

[FIC 16] FICHET-CALVET E., ÖLSCHLÄGER S., STRECKER T. *et al.*, "Spatial and temporal evolution of Lassa virus in the natural host population in Upper Guinea", *Scientific Reports*, vol. 6, p. e21977, 2016.

[FIS 01] FISHER-HOCH S.P., MC CORMICK J.B., "Towards a human Lassa fever vaccine", *Reviews in Medical Virology*, vol. 11, pp. 331–341, 2001.

[FRA 70] FRAME J.D., BALDWIN J.M., GOCKE D.J. *et al.*, "Lassa fever a new virus disease of man from West Africa, Clinical description and pathological findings", *American Journal of Tropical Medicine and Hygiene,* vol. 19, pp. 670–679, 1970.

[FRA 74] FRASER D.W., CAMPBELL C.C., MONATH T.P. *et al.*, "Lassa fever in the eastern Province of Sierra Leone, 1970–1972. Epidemiological aspects", *American Journal of Tropical Medicine and Hygiene*, vol. 23, pp. 1131–1139, 1974.

[GRA 86] GRANJON L. DUPLANTIER J.M., CASSAING J., "Etudes des relations sociales dans plusieurs populations du genre Mastomys (Rongeur, Muridé) au Sénégal: implications évolutives", *Collections Nationales du CNRS Biologie des Populations*, vol. 1, pp. 628–633, 1986.

[GRA 97] GRANJON L., DUPLANTIER J.M., CATALAN J. *et al.*, "Systematics of the genus Mastomys (Thomas, 1915) (Rodentia: Muridae). A review", *Belgian Journal of Zoology*, vol. 127, no. 1, pp. 7–18, 1997.

[GRA 05] GRANJON L., COSSON JF., QUESSEVEUR E. *et al.*, "Population dynamics of the multimammate rat Mastomys huberti in an annually flooded agricultural region of Central Mali", *Journal of Mammalogy*, vol. 86, no. 5, pp. 997–1008, 2005.

[GÜN 09] GÜNTHER S., HOOFD G., CHARREL R. *et al.*, "Mopeia virus-related arenavirus in natal multimammate mice, morogoro, Tanzania", *Emerg Infect Dis.*, vol. 15, no. 12, pp. 2008–2010, 2009.

[HAT 97] HATALSKI C.G., LIPKIN W.I., "Behavioral abnormalities and disease caused by viral infections of the central nervous system", in BECKAGE N.E. (ed.), *Parasites and Pathogens: Effects on Host Hormones and Behavior*, Springer, Berlin, 1997.

[HOL 01] HOLLAND C.V., COX D.M., "Toxocara in the mouse: a model for parasite-altered host behaviour?", *Journal of Helminthology*, vol. 75, pp. 125–135, 2001.

[HOL 06] HOLT J., DAVIS S., LEIRS H., "A model of Leptospirosis infection in an African rodent to determine risk to humans: Seasonal fluctuations and the impact of rodent control", *Acta Tropica*, vol. 99, no. 2, pp. 218–225, 2006.

[HUB 81] HUBERT B., COUTURIER G., POULET A. *et al.*, "Les conséquences d'un supplément alimentaire sur la dynamique des populations de rongeurs au Sénégal. I. Cas de Mastomys erythroleucus en zone sahélo-soudanienne", *Revue d'Ecologie Terre et Vie*, vol. 35, pp. 73–95, 1981.

[JOH 70] JOHNSON R.T., "The pathogenesis of experimental rabies", in NAGANO Y., DAVENPORT E.M. (eds), *Rabies*, University Park Press, Baltimore, 1970.

[KEA 77] KEANE E., GILLES H.M., "Lassa fever in Panguma Hospital, Sierra Leone, 1973–1976", *British Medical Journal*, vol. 1, pp. 1399–1402, 1977.

[KLE 02] KLEIN S.L., BIRD B.H., NELSON R.J. *et al.*, "Environmental and physiological factos associated with Seoul virus infection among urban populations of Norway rats", *Journal of Mammalogy*, vol. 83, pp. 478–488, 2002.

[KLE 03] KLEIN S.L., "Parasite manipulation of the proximate mechanisms that mediate social behavior in vertebrates", *Physiology & Behavior*, vol. 79, pp. 441–449, 2003.

[KLE 04] KLEIN S.L., ZINK M.C., GLASS G.E., "Seoul virus infection increases aggressive behaviour in male Norway rats", *Animal Behaviour*, vol. 67, pp. 421–429, 2004.

[LAL 08] LALIS A., Variabilité phénotypique et génétique des Mastomys (Rodentia, Muridae) de Guinée : évolution, environnement et infection virale, PhD thesis, UMPC, Paris, 2008.

[LAL 12] LALIS A., LEBLOIS R., LECOMPTE E. *et al.*, "The impact of human conflict on the genetics of mastomys natalensis and Lassa Virus in West Africa", *Plos One*, vol. 7, no.5, p. e37068, 2012.

[LAL 15] LALIS A., EVIN A., JANIER M. *et al.*, "Host evolution in Mastomys natalensis (Rodentia: Muridae): an integrative approach using geometric morphometrics and genetics", *Integrative Zoology*, vol. 10, no. 6, pp. 05–514, 2015.

[LAN 07] LANCASTER K., DIETZ D.M., MORAN T.H. *et al.*, "Abnormal social behaviors in young and adults rats neonatally infected with Borna disease virus", *Behavioural Brain Research*, vol. 176, pp. 141–148, 2007.

[LAV 98] LAVRENCHENKO L.A., LIKHNOVA O., BASKEVICH M.I. *et al.*, "Systematics and distribution of Mastomys (Muridae, Rodentia) from Ethiopia, with the description of a new species", *Zeitschrift für Säugetierkunde*, vol. 63, pp. 37–51, 1998.

[LEC 06] LECOMPTE E., FICHET-CALVET E., DAFFIS S. *et al.*, "Mastomys natalensis and Lassa fever, West Africa", *Emerging Infectious Diseases*, vol. 12, no. 12, pp. 1971–1974, 2006.

[LEC 07] LECOMPTE E., TER MEULEN J., EMONET S. *et al.*, "Genetic identification of Kodoko virus, a novel arenavirus of the African pigmy mouse (Mus Nannomys minutoides) in West Africa", *Virology,* vol. 364, no. 1, pp. 178–183, 2007.

[LEI 95] LEIRS H., Population Ecology of Mastomys natalensis (Smith, 1834). Implications for rodent control in Africa, PhD thesis, University of Antwerp, Antwerp, 1995.

[LEI 97] LEIRS H., STENSETH N.C., NICHOLS J.D. *et al.*, "Seasonality and non-linear density-dependence in the dynamics of African Mastomys rats", *Nature*, vol. 389, pp. 176–180, 1997.

[LIA 92] LIAO B.S., BYL F.M., ADOUR K.K., "Audiometric comparison of Lassa fever hearing loss and idiopathic sudden hearing loss: evidence for viral cause", *Otolaryngol Head Neck Surg*, vol. 106, pp. 226–229, 1992.

[LUK 93] LUKASHEVICH L.S., CLEGG J.C., SIDIBE K., "Lassa virus activity in Guinea: distribution of human antiviral antibody defined using enzyme-linked immunosorbent assay with recombinant antigen", *Journal of Medical Virology*, vol. 40, pp. 210–217, 1997.

[MAY 63] MAYR E., *Animal Species and Evolution*, Harvard University Press, Cambridge, 1963.

[MCO 87] MCCORMICK J.B., WEBB P.A., KREBS J.W. *et al.*, "A prospective study of the epidemiology and ecology of Lassa fever", *Journal of Infectious Diseases*, vol. 155, pp. 437–444, 1987.

[MER 73] MERTENS P.E., PATTON R., BAUM J.J. *et al.*, "Clinical presentation of Lassa fever cases during the hospital epidemic at Zorzor, Liberia, March-April 1972", *American Journal of Tropical Medicine and Hygiene*, vol. 22, pp. 780–784, 1973.

[MON 74] MONATH T.P., NEWHOUSE V.F., KEMP G.E. *et al.*, "Lassa virus isolation from Mastomys natalensis rodents during an epidemic in Sierra Leone", *Science,* vol. 185, pp. 263–265, 1974.

[MON 75] MONATH T.P., "Lassa fever: review of epidemiology and epizootiology", *Bulletin of the World Health Organization*, vol. 52, pp. 577–592, 1975.

[MON 84] MONSON M.H., FRAME J.D., JAHRLING P.B. *et al.*, "Endemic Lassa fever in Liberia, Clinical aspects at the Curran Lutheran Hospital, Zorzor, Liberia", *Transactions of the Royal Society of Tropical Medicine and Hygiene*, vol. 78, pp. 549–553, 1984.

[MOO 02] MOORE J., *Parasites and the Behavior of Animals*, Oxford University Press, Oxford, 2002.

[MUS 05] MUSSER G.G., CARLETON M.D., "Superfamily muroidea", in WILSON D.E., REEDER D.M. (eds), *Mammal Species of the World. A taxonomic and geographic reference*, 3rd Edition, John Hopkins University Press, Baltimore, 2005.

[MYL 15] MYLNE A.Q., PIGOTT D.M., LONGBOTTOM J. *et al.*, "Mapping the zoonotic niche of Lassa fever in Africa", *Transactions of the Royal Society of Tropical Medicine and Hygiene*, vol. 109, no. 8, pp. 483–492, 2015.

[MWA 02] MWANJABE P.S., SIRIMA F.B., LUSINGU J., "Crop losses due to outbreaks of Mastomys natalensis (Muridae, Rodentia) in the Lindi Region of Tanzania", *International Biodeterioration & Biodegradation*, vol. 49, pp. 133–137, 2002.

[OLA 16] OLAYEMI A., CADAR D., MAGASSOUBA N. *et al.*, "New hosts of the lassa virus", *Scientific Reports*, vol. 6, p. e25280, 2016.

[PET 77] PETTER F., "Les rats à mamelles multiples d'Afrique occidentale et centrale: Mastomys erythroleucus (Temminck, 853) et Mastomys huberti (Wroughton, 1908)", *Mammalia,* vol. 41, no. 4, pp. 441–444, 1977.

[RAM 03] RAMBUKKANA A., KUNZ S., MIN J. *et al.*, "Targeting Schwann cells by nonlytic arenaviral infection selectively inhibits myelination", *Proceedings of the National Academy of Sciences*, vol. 100, no. 26, pp. 16071–16076, 2003.

[RIC 03] RICHMOND J.K., BAGLOLE D.J., "Lassa fever: epidemiology, clinical features, and social consequences", *British Medical Journal*, vol. 327, pp. 12751–12755, 2003.

[SHA 02] SHARMA A., VALADI N., MILLER A.H. *et al.*, "Neonatal viral infection decreases neuronal progenitors and impairs adult neurogenesis in the hippocampus", *Neurobiology of Disease*, vol. 11, pp. 246–256, 2002.

[SOL 95] SOLBRIG M.V., FALLON J.H., LIPKIN W.I. *et al.*, "Behavioral disturbances and pharmacology of Borna disease", *Current Topics in Microbiology and Immunology*, vol. 190, pp. 93–101, 1995.

[TER 96] TER MEULEN J., LUKASHEVICH I., SIDIBE K. *et al.*, "Hunting of peridomestic rodents and consumption of their meat as possible risk factors for rodent-to-human transmission of Lassa virus in the Republic of Guinea", *American Journal of Tropical Medicine and Hygiene*, vol. 55, pp. 661–666, 1996.

[TOM 88] TOMORI O., FABIYI A., SORUNGBE A. *et al.*, "Viral hemorrhagic fever antibodies in Nigerian populations", *American Journal of Tropical Medicine and Hygiene*, vol. 38, pp. 407–410, 1988.

[VOL 01] VOLOBOUEV V., HOFFMAN A., SICARD B. *et al.*, "Polymorphism and polytypy for pericentric inversion in 38-chromosome Mastomys (Rodentia, Murinae) and possible taxonomic implications", *Cytogenetics & Cell Genetics*, vol. 92, pp. 237–242, 2001.

[WHI 72] WHITE H.A., "Lassa fever: a study of 23 hospital cases", *Transactions of the Royal Society of Tropical Medicine and Hygiene*, vol. 66, pp. 390–401, 1972.

[WOR 16] WORLD HEALTH ORGANIZATION, Lassa fever sensitization campaigns, report, 2016.

[WOR 17] WORLD HEALTH ORGANIZATION, Lassa fever, report, 2017.

[WRI 95] WRIGHT K.E., AHMED R., BUCHMEIER M.J., "Persistent infection of mice with Pichinde virus associated with failure to thrive", *Microbial Pathogenesis*, vol. 19, pp. 73–82, 1995.

12

Evolutionary History of Moles in Western Europe: One Mole May Hide Another!

12.1. An unexpected biodiversity

If there's one mammal all garden owners know well, without ever having actually seen one, it is the mole. A friendly character in childhood cartoons (*The Mole* by Zdenek Miler) or comics (*Les Riquiquis* from the French magazine *Youpi j'ai compris*), this animal is often hated by gardeners and lawn tenders who hardly appreciate the numerous molehills (Figure 12.1) left behind by those that dig up the earth when making their underground tunnels. The methods put into practice to get rid of them in a more or less natural way, or to kill them, are therefore numerous: fritillary or Euphorbia bulbs, ultrasonic mole repellers, putting dog hair in the molehills, mole-killing chemical products, mole-traps, fumigation, etc. not to mention professional mole-trappers! There is in fact a whole network of professional mole-trappers who are experts in the art of using traditional mole traps, and have followed theoretical and practical training courses for one to three years [DOR 12].

Chapter written by Violaine NICOLAS, Jessica MARTINEZ-VARGAS and Jean-Pierre HUGOT.

Figure 12.1. *Photo of molehills in a garden*

But what do we really know about moles? In this chapter, we will focus on moles from the genus *Talpa*. All the species of this genus look alike morphologically, are fossorial and live in underground burrows. Moles have long been considered members of the mammalian order Insectivora, alongside other mammals that feed on insects such as hedgehogs, shrews, golden moles, tenrecs or solenodons. However, current classifications no longer recognize the Insectivora as an order, and moles are now considered as part of the Soricomorpha (together with the solenodons, shrews and a family of mammals now extinct: the nesophontes). Within the Soricomorpha, moles of the genus *Talpa* belong to the family Talpidae and to the sub-family Talpinae. The genus *Talpa* is widely distributed throughout the whole Western Palearctic region, from the Iberian Peninsula to Siberia [HUT 05]. According to the most recent systematic revision of this genus, nine species are currently recognized [HUT 05] and form a monophyletic group with a common ancestor [COL 10, BAN 15]: the Siberian mole *T. altaica*, Nikolasky, 1883; the Caucasian mole *T. caucasica*, Satunin, 1908; the Levant mole *T. levantis*, Thomas, 1906; Père David's mole, *T. davidiana*, Milne-Edwards, 1884; the blind mole *T. caeca*, Savi, 1882; the European mole *T. europaea*, Linnaeus, 1758; the Iberian mole *T. occidentalis*, Cabrera, 1907; the Roman mole *T. romana*, Thomas, 1965; and the Balkan mole *T. stankovici*, Martino and Martino, 1931. Based on the genetic data, Bannikova *et al.* [15] recently recognized three additional species: *T. talyschensis*, Vereschagin, 1945; *T. ognevi*, Stroganov, 1948 and

Talpa ex gr. levantis. How can we explain the discovery of new species within a widespread, abundant and easy to capture mammalian group? An explanation could be the fact that the morphology of the species included in the genus *Talpa* is very constrained by their underground lifestyle which results in an apparent morpho-anatomical similarity [NIE 90, KRY 01].

It is interesting to note that all mole species, except *T. europaea*, are endemic to small geographical areas [MIT 99, HUT 05]. Thus, until very recently it was thought that only the European mole was widely distributed in Europe: from the Ebro River in Spain to the Ob and Irtysh Rivers in Russia [HUT 05]. Despite its abundance in its natural habitat, the European mole has been scarcely studied and little data are available on its genetic and morphological variability. Up to 26 different forms were described, but only three sub-species are currently recognized [HUT 05]: *europaea*, *cinerea* and *velessiensis*. Loy & Corti [LOY 96] showed wide variation in the morphology of the mandible of *T. europaea*, related to both geography and climate. They also showed distinct morphologies in peripheral populations (Italian, Spanish and British) with respect to those from the rest of Europe. A recent study on the genetic variability of this species has shown the existence of three distinct mitochondrial lineages: one endemic to Northern Spain, one endemic to Italy and one widely spread throughout the rest of Europe [FEU 15]. The most surprising in this study is that the Spanish lineage forms a monophyletic group with the Iberian mole, *T. occidentalis*, and not with the other two European mole lineages. This is an important point as it may suggest the existence of a new species in Spain!

12.2. A new mole species in France and Spain?

The fact that specimens of *T. europaea* from Northern Spain are, in terms of mitochondrial DNA, genetically closer to *T. occidentalis* than to *T. europaea* from Italy or from the rest of Europe suggests:

– either the existence of a new species in Spain,

– or that the distribution range of the species *T. europaea* and *T. occidentalis* must be modified (specimens from Northern Spain would then belong to the species *T. occidentalis*, not to *T. europaea*),

– or a past phenomenon of introgression (i.e. transfer) of mitochondrial DNA from the Iberian mole to the European mole at the end of the Pliocene era, following processes of hybridization.

In order to test these hypotheses and better define the geographical distribution of the different mole lineages, we have sequenced the mitochondrial and nuclear DNA of a large number of moles of these two species throughout their distribution range, and we have completed our study by a morphological study of the sequenced specimens [NIC 15, NIC 17a].

12.2.1. *Mitochondrial DNA: three lineages of* T. europaea

Based on mitochondrial DNA sequences (Cytochrome b gene) of 396 individuals, we have been able to show that the specimens previously identified as *T. europaea* include three main lineages (Figures 12.2 and 12.3): a European lineage, an Italian lineage and a Spanish–French lineage. Furthermore, *T. europaea* appears paraphyletic because the Spanish–French lineage is phylogenetically closer to *T. occidentalis* than to *T. europaea*. The genetic divergence between this lineage and the three others varies from 7 to 8%. The Spanish–French lineage would have split from *T. occidentalis* at the very beginning of the Pleistocene, around 2.47 ± 0.12 million years ago.

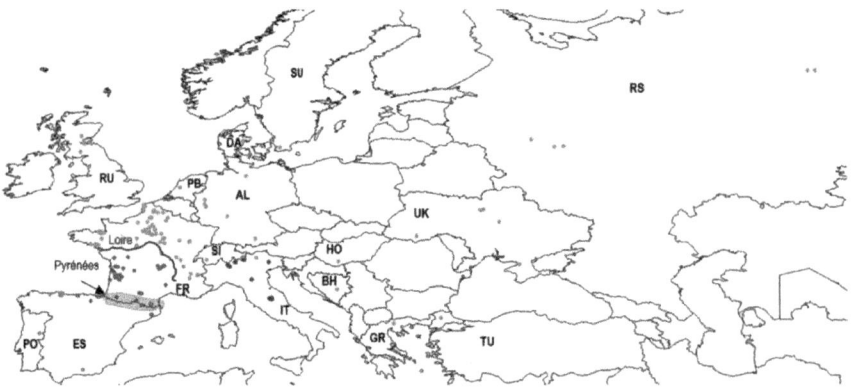

Figure 12.2. *Map showing the collection points of specimens included in the genetic analysis and the main phylogenetic lineages identified by mitochondrial DNA (Cytochrome b gene):* T. occidentalis *(orange) and* T. europaea *(pink: European lineage, violet: Italian lineage, green: Spanish–French lineage). For a color version of this figure, see www.iste.co.uk/grandcolas/biodiversity.zip*

The Spanish–French lineage includes all the specimens previously identified as *T. europaea* from Spain and all the specimens from Southwestern France, with the exception of two specimens from Mosset in the Pyrénées-Orientales French department. In France, it has not been identified to the north or east of the Loire River.

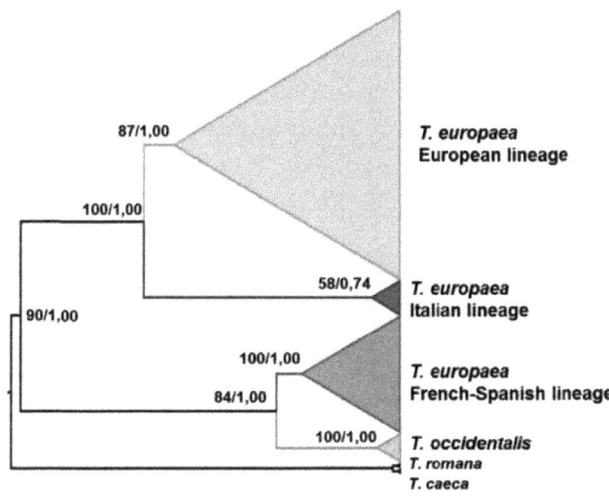

Figure 12.3. *Phylogeny recovered by the maximum-likelihood analysis (model GTR+I+G) from the sequences of mitochondrial DNA (Cytochrome b gene). Numbers on nodes represent the ML bootstrap support values and the Bayesian posterior probabilities, respectively. For a color version of this figure, see www.iste.co.uk/ grandcolas/biodiversity.zip*

The Italian lineage comprises all the Italian individuals, except one specimen from Bolzano.

The European lineage includes all specimens from Bosnia and Herzegovina, Denmark, Germany, Greece, Hungary, the Netherlands, Russia, Switzerland, Sweden, Ukraine, the UK, Turkey and Northeastern France (north and east of the Loire, plus two specimens from the Pyrenees).

The genetic divergence between the Italian and European lineages is a bit less than 2%, which corresponds to a divergence around 0.667 ± 0.115 million years ago.

12.2.2. *Nuclear DNA:* T. europaea *is paraphyletic*

A nuclear gene, the intron 10 of the histone deacetylase 2, was sequenced for 95 specimens that represented the main mitochondrial lineages:

T. occidentalis, the Spanish–French lineage and the European lineage of *T. europaea*. For technical reasons, we have not been able to obtain sequences for the Italian lineage of *T. europaea*. As the genetic variation obtained for this gene was quite low (18 polymorphic sites), we will present the results obtained in the form of a network of haplotypes (Figure 12.4), each haplotype corresponding to a unique DNA sequence different from the other sequences by at least one mutation. All the sequences from *T. occidentalis* are identical and correspond to a single haplotype. Within *T. europaea*, two groups of haplotypes can be identified: one corresponding to the European mitochondrial lineage, and the other to the Spanish–French mitochondrial lineage. In agreement with the mitochondrial data, the genetic divergence between the Spanish–French and the European lineages is higher (more than 12 mutations) than observed between the Spanish–French lineage and *T. occidentalis* (1 to 3 mutations). Haplotypes of the European lineage differ by more than 11 mutations from haplotypes of the *T. occidentalis* lineage.

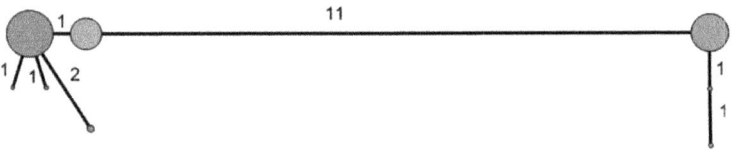

Figure 12.4. *Haplotype network obtained from nuclear sequences (intron 10 of the histone deacetylase 2). The length of the branches is proportional to the number of substitutions, and the figures indicate the number of substitutions between haplotypes. The circles represent haplotypes (unique DNA sequences) and the size of the circles is proportional to the number of individuals carrying that haplotype. The colors represent the different mitochondrial lineages: pink: European lineage of* T. europaea, *green: Spanish–French lineage of* T. europaea *and orange:* T. occidentalis. *Adapted from Nicolas et al. [NIC 17a]. For a color version of this figure, see www.iste.co.uk/grandcolas/biodiversity.zip*

A group is said to be paraphyletic when it includes an ancestor but not all of its descendants. As the paraphyly of *T. europaea* is revealed by both mitochondrial and nuclear genetic data, our results allow us to invalidate the hypothesis of a past phenomenon of introgression of mitochondrial DNA from the Iberian mole into the European mole at the end of the Pliocene period. The populations of *T. europaea* from Spain and Southwestern France must be considered either as belonging to *T. occidentalis* or to a not yet described new species. In order to decide between these two hypotheses, it may be useful to use genetic distance between taxa as a basis. According to

Colangelo *et al.* [COL 10], the genetic distance between species from the *Talpa* genus varies, for the Cytochrome b gene, between 8% and 15%. As these values are similar to those observed between the Spanish–French lineage and the two other lineages of *T. europaea*, and close to that observed between the Spanish–French lineage and *T. occidentalis* (7%), our genetic data suggest that a specific status could be granted to the Spanish–French lineage. Nonetheless, using mtDNA genetic distances to assign taxonomic rank [JOH 98, BAK 06] is problematic and is not without its detractors [FER 02]. Another way to arrive at a taxonomical conclusion is to compare the morphology of these populations: they may be considered as different taxonomical unities if diagnostic traits may be identified.

12.2.3. Morphological analysis: a new mole species

Talpa europaea differs from *T. occidentalis* by its open eyes: in *T. occidentalis*, the eyelids are fused together and completely covered by membranes, while in *T. europaea*, the eyes, although very small, are open [PET 71, NIE 90, WIT 97, AUL 08]. All specimens from the Spanish–French lineage have their eyelids fused together, as observed in *T. occidentalis*.

The species *T. europaea* and *T. occidentalis* can also be distinguished by their external body measurements: *T. europaea* is larger than *T. occidentalis* [MIL 12, JIM 84, NIE 90, AUL 08]. Our results show that weight, body size and foot size are significantly larger in the specimens from the Spanish–French lineage (weight = 89 ± 17 g, body = 149 ± 7 mm, foot = 21.5 ± 1.5 mm) than in those from the European lineage (weight = 76 ± 12 g, body = 144 ± 8 mm, foot = 20.9 ± 1.6 mm; P < 0.02 for the three measurements according to the students' t-test).

The last trait often used to differentiate *T. europaea* and *T. occidentalis* concerns the mesostyle of the first upper molar (M1), which is simple in *T. europaea* and double in *T. occidentalis* [MIL 12, CAP 81, NIE 90, CLE 01, AUL 08]. All specimens from the Spanish–French lineage have a simple M1 mesostyle, as in *T. europaea* (Figure 12.5). In *T. europaea* and *T. occidentalis*, the mesostyle of the second (M2) and the third (M3) upper molar is divided into two cusps of similar size, aligned on the same plane. Conversely, the specimens from the Spanish–French lineage have a simple M2 mesostyle, and a M3 mesostyle consisting of a large anterior cusp and a smaller posterior cusp, so small that it can be difficult to see in some individuals.

T. occidentalis

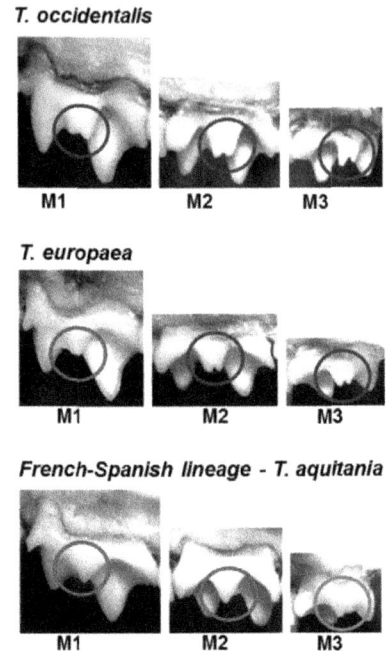

T. europaea

French-Spanish lineage - T. aquitania

Figure 12.5. *Mesostyles of the three upper molars (from left to right: M1, M2, M3), labial view, of* T. occidentalis, T. europaea *and of the Spanish–French lineage. The color of the circles corresponds to the different types of mesostyles: simple mesostyle in red, mesostyle with two cuspids of similar size aligned on the same plane in blue and mesostyle with unequal cusps in green. Adapted from Nicolas et al. [NIC 17a]. For a color version of this figure, see www.iste.co.uk/grandcolas/biodiversity.zip*

In conclusion, the Spanish–French lineage has both original characteristics (e.g. mesostyle of M2 and M3) and characters observed in one or the other of the two closest species. Some characters more closely resemble *T. occidentalis* (e.g. eyes) and others *T. europaea* (e.g. external measurements and mesostyle of M1). These results, combined with our genetic data, suggest that the Spanish–French lineage must be recognized as a new species. We have proposed to name it *T. aquitania* [NIC 15, NIC 17b].

12.3. Factors affecting the geographical distribution of the species in France

Based on our genetic data, *T. aquitania* and *T. europaea* have an allopatric distribution on both sides of the Loire River (apart from two specimens of *T. europaea* captured within the distribution range of the *T. aquitania* in the Eastern Pyrenees). The same distribution pattern, with two lineages on both sides of the Loire River, can be observed in other small mammals known for digging burrows, like the common vole [TOU 08].

How to explain this distribution pattern? The geographical distribution of a given species is often constrained by climatic variables. Temperature and precipitation affect the hardness of the ground (through freezing, drought and floods), which can have an influence on the ability of moles to dig, and also on the availability of their prey, small invertebrates in the ground [GOR 90]. We built a species distribution model for each of the two French species based on the presence localities of both taxa. As environmental layers, we used the available climatic data from the WorldClim database (http://www.worldclim.org/). Our model shows that the south of France, from the edge of the Alps to Eastern Pyrenees, is favorable to *T. europaea* (Figure 12.6). It is therefore not surprising to have captured this species in Mosset (in the department of Pyrénées-Orientales). According to our Species Distribution Modeling analyses, *T. europaea* should also be captured around the Gironde estuary; however, despite a particularly intensive sampling, we never have captured any specimen in this region. Regarding *T. aquitania*, and under present-day bioclimatic conditions, the climatic model revealed a high habitat suitability across northern Spain and most of France (except the northeast). This area is larger than the current known distribution of the species. Thus, factors other than climatic ones probably explain the geographical distribution of these two species. Our model does not take into account interactions between species (e.g. predation, competition, parasitism and mutualism) and this is important since realized versus fundamental niches often differ [PEZ 08, SIN 10, GIA 13]. The fundamental niche can be defined as the range of environmental conditions in which a species may survive, whereas the realized niche can be defined as the range of environmental conditions in which a species is actually found. Mutual competitive exclusion between the two French species of mole could explain their allopatric geographic distribution. In other subterranean mammals, like the pocket gopher (a rodent), interspecific differences in body size and digging strategy have been shown to confer competitive dominance of one species over another, depending on soil characteristics [MAR 13]. In regions where different soil types occur, several species of pocket gopher can coexist [THA 68]. The same thing has been observed in Japanese moles from the *Mogera* genus, where the dominant species, *Mogera wogura*, is progressively expanding its range northwards, displacing the smaller species,

M. imaizumii. The hardness of the soil affects the geographic distribution of these species and allows their coexistence in a very specific set of conditions [ABE 01]. Since there are significant differences in size between the two French species of mole, it would be interesting to study how these differences, as well as the moles' methods of digging, affect their dominance in cases of competition. Furthermore, it would also be important to take into account the soil's characteristics (clay percentage, density, hardness, etc.) in the niche modelization of these species. Do the interspecific differences in body size and digging method explain the allopatric distribution of *T. aquitania* or *T. europaea* over most of the French territory, but their coexistence in the Eastern Pyrenees?

T. europaea

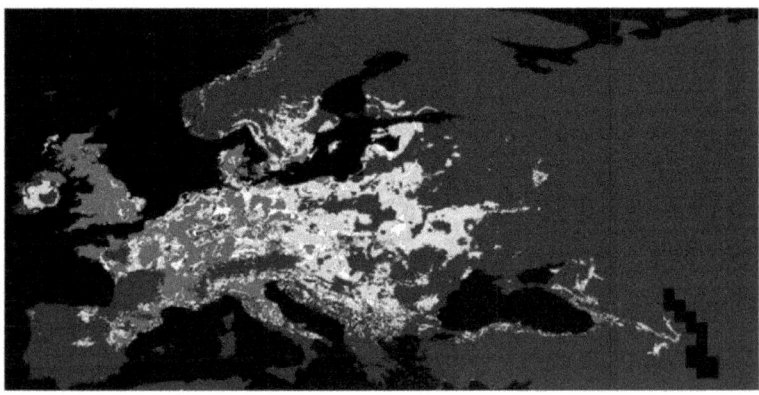

T. aquitania

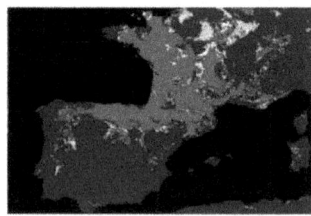

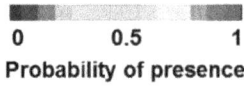

0   0.5   1
Probability of presence

Figure 12.6. *Species distribution modeling of* T. europaea *and* T. aquitania *estimated by Maxent from 11 bioclimatic variables from the database WorldClim [BIO1, BIO2, BIO3, BIO7, BIO9, BIO10, BIO14, BIO15, BIO16, BIO18, BIO19]. Warmer colors show areas with higher probability of presence. For a color version of this figure, see www.iste.co.uk/grandcolas/biodiversity.zip*

The Loire River alone does not seem sufficient to explain the allopatric distribution of these two species, as moles can dig tunnels that may lie anywhere from 5 to 150 cm below the surface; moles can also swim for 30–50 minutes covering a distance of over one kilometer [GOR 90]. Furthermore, the genetic data show that other large rivers such as the Seine, the Garonne or the Rhin do not constitute effective barriers to the spread of moles [NIC 17a]. To better understand the actual role that the Loire River may play as a barrier to the spread of moles, it would be pertinent to carry out a targeted sampling of the two shores of this river from its source to its mouth. It is necessary to highlight here that the complex history [CYP 04, NEH 10, PRO 10] of the Loire throughout the Plio-Pleistocene (marine invasions, creation of a system of fluvial terraces) could have contributed to maintain the physical separation between the two mole species.

In conclusion, the genetic studies [BAN 15, FEU 15, NIC 17a] carried out during the last few years on moles in Europe allow us to highlight a hitherto unsuspected biodiversity. Thus, even in a biogeographic area reputed for being well studied, Europe, and for a common mammal species, there are still species left to be described. Inventory campaigns in the field, involving specimen collection, and the combination of genetic, morphological and cytogenetic studies must therefore be pursued in order to identify these cryptic (i.e. morphologically similar) species. Some will wonder why would that be so important? We answer that a precise taxonomic identification is of the utmost importance not only for zoological science and evolutionary research, but also for biodiversity management, conservation of endangered taxa, pest control and public health. Taxonomical studies must be accompanied by precise studies on the ecology of these species, in order to better define their ecological niche and trace their evolutionary history.

12.4. Bibliography

[ABE 01] ABE H., "Soil hardness, a factor affecting the range expansion of *Mogera wogura* in Japan", *Mammal Study*, vol. 26, pp. 45–52, 2001.

[AUL 08] AULAGNIER S., HAFFNER P., MITCHELL-JONES A.J. *et al.*, *Guide des mammifères d'Europe, d'Afrique du Nord et du Moyen-Orient*, Delachaux et Niestlé, Paris, 2008.

[BAK 06] BAKER R.J., BRADLEY R.D., "Speciation in mammals and the genetic species concept", *Journal of Mammalogy*, vol. 87, pp. 643–662, 2006.

[BAN 15] BANNIKOVA A.A., ZEMLEMEROVA E.D., COLANGELO P. *et al.*, "An underground burst of diversity – a new look at the phylogeny and taxonomy of the genus *Talpa* Linnaeus, 1758 [Mammalia: Talpidae] as revealed by nuclear and mitochondrial genes", *Zoological Journal of the Linnean Society*, vol. 175, pp. 930–948, 2015.

[CAP 81] CAPANNA E., "Karyotype and cranial morphology of *Talpa romana* [THOMAS] from terra typical", *Mammalia*, vol. 45, pp. 71–82, 1981.

[CLE 01] CLEEF-RODERS V.J.T., HOEK OSTENDE V.D.L.W., "Dental morphology of *Talpa europaea* and *Talpa occidentalis* [Mammalia: Insectivora] with a discussion of fossil Talpa in the Pleistocene of Europe", *Zoologische Mededelingen*, vol. 75, pp. 51–68, 2001.

[COL 10] COLANGELO P., BANNIKOVA A.A., KRYSTUFEK B. *et al.*, "Molecular systematics and evolutionary biogeography of the genus *Talpa* [Soricomorpha: Talpidae]", *Molecular phylogenetics and evolution*, vol. 55, pp. 372–380, 2010.

[CYP 04] CYPRIEN A.-L., VISSET L., CARCAUD N., "Evolution of vegetation landscapes during the Holocene in the central and downstream Loire basin [Western France]", *Vegetation History and Archaeobotany*, vol. 13, pp. 181–196, 2004.

[DOR 12] DORMION J., *Le piégeage traditionnel des taupes*, Editions Eugen Ulmer, Paris, 2012.

[FER 02] FERGUSON J.W., "On the use of genetic divergence for identifying species", *Biological Journal of the Linnean Society*, vol. 75, pp. 509–516, 2002.

[FEU 15] FEUDA R., BANNIKOVA A.A., ZEMLEMEROVA E.D. *et al.*, "Tracing the evolutionary history of the mole, *Talpa europaea*, through mitochondrial DNA phylogeography and species distribution modeling", *Biological Journal of the Linnean Society*, vol. 114, pp. 495–512, 2015.

[GIA 13] GIANNINI T.C., CHAPMAN D.S., SARAIVA A.M. *et al.*, "Improving species distribution models using biotic interactions: a case study of parasites, pollinators and plants", *Ecography*, vol. 36, pp. 649–656, 2013.

[GOR 90] GORMAN M.L., STONE R.D., *The Natural History of Moles*, Christopher Helm Publications, London, 1990.

[HUT 05] HUTTERER R., "Order Soricomorpha", in D.E. WILSON, D.M. REEDER (eds), *Mammal Species of the World: A Taxonomic and Geographic Reference*, 3rd ed., Johns Hopkins University Press, Baltimore, 2005.

[JIM 84] JIMÉNEZ R., BURGOS M., DIAZ DE LA GUARDIA R., "Karyotype and chromosome banding in the mole [*Talpa occidentalis*] from the south-east of the Iberian Peninsula. Implication on its taxonomic position", *Caryologia: International Journal of Cytology*, vol. 37, pp. 253–258, 1984.

[JOH 98] JOHNS G.C., AVISE J.C., "A comparative summary of genetic distances in the vertebrates from the mitochondrial cytochrome *b* gene", *Molecular Biology and Evolution*, vol. 15, pp. 1481–1490, 1998.

[KRY 01] KRYŠTUFEK B., VOHRALÍK V., *Mammals of Turkey and Cyprus. Introduction checklist Insectivora*, Science and Research Centre of the Republic of Slovenia, Koper, 2001.

[LOY 96] LOY A., CORTI M., "Distribution of *Talpa europea* [Mammalia, Insectivora, Talpidae] in Europe: A biogeographic hypothesis based on morphometric data", *Italian Journal of Zoology*, vol. 63, pp. 277–284, 1996.

[MAR 13] MARCY A.E., FENDORF S., PATTON J.L. *et al.*, "Morphological adaptations for digging and climate-impacted soil properties define Pocket Gopher [*Thomomys* spp.] Distributions", *Plos One*, vol. 8, p. e64935, 2013.

[MIL 12] MILLER G.S., Catalogue of the mammals of Western Europe [Europe exclusive of Russia] in the collection of the British Museum, Trustees of the British Museum, London, 1912.

[MIT 99] MITCHELL-JONES A.J., AMORI G., BOGDANOWICZ W. *et al.*, *The Atlas of European Mammals*, T & AD Poyser Natural History, London, 1999.

[NEH 10] NEHLIG P., "Géologie du bassin de la Loire", *Géosciences*, vol. 12, pp. 10–23, 2010.

[NIC 15] NICOLAS V., MARTÍNEZ-VARGAS J., HUGOT J.-P., "*Talpa aquitania* nov. sp. (Talpidae, Soricomorpha) a new mole species from southwest France and north Spain", *Bulletin de l'Académie vétérinaire de France*, vol. 166, no. 4, pp. 1–6, 2015.

[NIC 17a] NICOLAS V., MARTÍNEZ-VARGAS J., HUGOT J.-P., "Molecular data and ecological niche modelling reveal the evolutionary history of the common and Iberian moles [Talpidae] in Europe", *Zoologica Scripta*, vol. 46, pp. 12–26, 2017.

[NIC 17b] NICOLAS V., MARTÍNEZ-VARGAS J., HUGOT J.-P., "*Talpa aquitania* nov. sp. [Talpidae, Soricomorpha] a new mole species from southwest France and north Spain", *Mammalia*, vol. 81, no. 6, pp. 642–642, 2017.

[NIE 90] NIETHAMMER J., KRAPP F., *Handbuch der Säugetiere Europas, 3/I: Insektenfresser, Primaten*, Aula-Verlag, Wiesbaden, 1990.

[PEA 08] PEARMAN P.B., GUISAN A., BROENNIMANN O. *et al.*, "Niche dynamics in space and time", *Trends in Ecology & Evolution*, vol. 23, pp. 149–158, 2008.

[PET 71] PETROV B.M., "Taxonomy and distribution of moles [genus *Talpa*, Mammalia] in Macedonia", *Acta Musei Macedonici Scientiarum Naturalium*, vol. 12, pp. 117–138, 1971.

[PRO 10] PROUST J.-N., RENAULT M., GUENNOC P. *et al.*, "Sedimentary architecture of the Loire River drowned valleys of the French Atlantic shelf", *Bulletin de la Société Géologique de France*, vol. 181, pp. 129–149, 2010.

[SIN 10] SINCLAIR S.J., WHITE M.D., NEWELL G.R., "How useful are species distribution models for managing biodiversity under future climates?", *Ecology and Society*, vol. 15, p. 8, 2010.

[THA 68] THAELER C.S., *An Analysis of the Distribution of Pocket Gopher Species in Northeastern California [Genus Thomomys]*, University of California Press, Berkeley, 1968.

[TOU 08] TOUGARD C., RENVOISE E., PETITJEAN A. *et al.*, "New insight into the colonization processes of common voles: inferences from molecular and fossil evidence", *PloS One*, vol. 3, p. e3532, 2008.

[WIT 97] WITTE G.R., *Der Maulwurf Talpa europaea*, Westarp Wissenschaften, Magdeburg, 1997.

The Conoidea and Their Toxins: Evolution of a Hyper-Diversified Group

13.1. General introduction and state of the art

Conoidea have been attracting the interest of systematicians and toxinologists for several decades. Systematicians first, because this group of marine gastropods is extremely rich in species, with a huge variation in size, shape and patterns unique in mollusks. This great diversity, associated with a high variability of shell, explains why the successive attempts to understand their species diversity and phylogenetic relationships have proved unfruitful for the most part until very recently. Within the Conoidea superfamily, the turrids, a non-natural group, as we shall see later, have been considered for nearly two centuries as a taxonomic nightmare: "In no other group of mollusks is it so difficult to make a satisfactory classification as in the Pleurotomidae [=Turridae]" [TRY 84]. On the contrary, cone snails, which are easier to collect and for the most part large, with a diversity proportionally smaller than turrids, make up the most collected group of shellfish, and their alpha taxonomy is the most stable of all the Conoidea. Whether they be relatively well known, like cone snails, or a complete mystery, like turrids, the systematicians have their work cut out for them with this vast array of little-known mollusks, which are in fact more diverse than birds and perhaps even more diverse than birds and mammals together.

Toxinologists are attracted by Conoidea as they are venomous predators. They prey on worms, other mollusks and even fish, with a combination of

Chapter written by Nicolas PUILLANDRE and Sébastien DUTERTRE.

toxins they produce in their venom gland, which they inject into their prey through a specialized venom delivery system, the structure of which resembles a hypodermic needle in some species. These toxins, which have been studied since the 1970s, act for the most part upon the transmission of neuronal signals, by targeting membrane receptors in a very specific way. They can therefore be used as a model to understand the physiological and molecular mechanisms of nervous signaling, but they have also proved useful in therapeutic treatments for humans.

In this chapter, we will start by presenting the latest developments in the systematics and toxinology of Conoidea at the beginning of the 21st Century. We will then move on to recent developments linked to upcoming molecular approaches in systematics and "omics" approaches in toxinology, particularly linked to the revolution of the NGS. We will conclude by defending the idea, with several examples, that an understanding of the evolutionary processes behind the extraordinary diversity of Conoidea and their toxins can only be reached by integrating the results of different disciplines, namely systematics and toxinology, supplemented with ecological, behavioral and even physiological data.

13.1.1. *Systematics*

Until the 1990s, the systematics of Conoidea, whether they concerned the delimitation of taxa (taxonomy) or the inference of their phylogenetic relationships, were essentially based on morphological characters, mainly of the shell and sometimes of the radula. Three families were generally recognized: cone-shaped Conoidea, Conidae; Conoidea with a very elongated shell, Terebridae; and the rest, that is to say, all those species belonging to neither the Conidae nor the Terebridae, in the Turridae family (="turrids" henceforth). Some authors have also proposed subdivisions into subfamilies [e.g. POW 66], but yet again, even though some species were allocated based on synapomorphies, the majority were allocated "by default", due to the absence of diagnostic traits. The same problem exists at the genus level, and only very few genera have passed the molecular test (see section 13.2.1).

As in other groups of gastropod, shell variability does not always reflect the limits between taxa, as different species, sometimes belonging to different genera or even different families, may have very similar shells, or on the contrary, although less common, conspecific specimens may show very different shells. Phenotypic plasticity, convergence or retention of ancestral characters may explain these patterns. However, in Conoidea, the problem is exacerbated by the very large number of species, the majority of which are still unknown, with an inter- and intra-specific variability that is seemingly infinite: "The forms are exceedingly numerous, and known in many species to be very variable in their characters" [TRY 84]. Among the numerous examples of taxa that are phylogenetically distant and yet morphologically very similar, let us cite the genera *Cochlespira* (Cochlespiridae) and *Toxicochlespira* (Mangeliidae), which diverged at least 60 million years ago, the genera *Leucosyrinx* (Pseudomelatomidae) and *Sibogasyrinx* (Cochlespiridae), also separated by at least 60 million years ago, or indeed the species *Strictispira paxillus* and *Crassispira fuscescens*, basically indiscernible and considered, morphologically, as conspecific by specialists of this group (Figure 13.1). These two species from the same family but different genera also have another particularity: one has conserved the venom gland typical of the Conoidea, but the other (*S. paxillus*) has subsequently lost it.

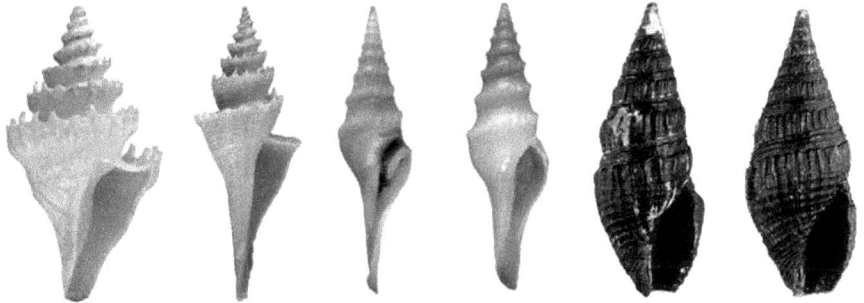

Figure 13.1. *From left to right:* Toxicochlespira *sp. (Mangeliidae),* Cochlespira *sp. (Cochlespiridae),* Sibogasyrinx *sp. (Cochlespiridae),* Leucosyrinx *sp. (Pseudomelatomidae),* Crassispira *sp. (Pseudomelatomidae) and* Strictispira paxillus *(Pseudomelatomidae)*

The integration of anatomical traits has somewhat disrupted traditional classifications and allowed us to recognize seven families within the Conoidea, with most turrids found to be spread among the Conidae and in several new families [TAY 93]. However, as for previous classifications, these families are defined phenetically and do not reflect, for the most part, phylogenetic relationships between taxa. For example, some species that have secondarily lost the radula, such as the species in the genus *Strictispira*, have been transferred to a separate family, the Strictispiridae, without it having been demonstrated that they really constitute a phylogenetically independent lineage. After 20 years, phylogenetic approaches based on molecular traits will change the classification yet again.

At the species level, analysis of morphologic diversity is equally subject to interpretation, and hypotheses regarding species are often more opinions than testable scientific hypotheses. Descriptions of species in which the morphologic traits are formalized are rare, and contradictions between specialists are common. Even in Conidae and to a slightly lesser extent in Terebridae, far better known than turrids, many species delimitations are not unanimously agreed upon by taxonomists, with some newly described species being synonymized some years later. Cone snails remain, however, the only group for which the number of species described by taxonomists has not decreased since 1900 [BAC 12]: the community of amateur taxonomists, for the most part shell collectors, is very active, and no fewer than 50 new species of cone snails were described in 2014. In total, more than 5,000 species of Conoidea are described, but it is thought that at least double that, if not triple, are yet to be described.

13.1.2. *Toxinology*

Venomous animals have always both frightened and attracted humankind, not only because of their capacity to bring down a fully grown man but also the pharmacological properties of their venom that can paradoxically cure certain illnesses. For example, we find in ancient Chinese pharmacopoeia the use of various concoctions made from venoms or various venomous animal extracts (snakes, scorpions, spiders, amphibians) to treat rheumatism or cancer. At present, there are approximately 20 types of medicine that have come directly from research on animal toxins, which allow us to treat chronic pain, imbalance of the hemostase or even diabetes. Among marine venomous animals, one can cite the magical cone (*Conus magus*) as an example, whose venom has made it possible to discover Prialt™ (see below), an analgesic more potent than morphine [MIL 04]. Cones (on which we will focus – see for example [BOU 11] for information on other Conoidea) are the only representatives of Conoidea that possess a venomous organ sufficiently robust to produce venom powerful enough to cause significant symptoms in humans. In fact, in most turrids and terebrids, the radula that allows the injection of venom barely measures longer than a few hundreds of microns, whereas in cones, the radula is modified into the shape of a harpoon and can reach more than 10 mm in length. The shape of this radula is often characteristic of the species' diet (Figure 13.2). The highly specialized venomous apparatus (Figure 13.2) is composed of a venom gland, a salivary gland (and sometimes auxiliary salivary glands), a radular sac, a pharynx, a proboscis and a muscular bulb. The radula is produced in the long arm of the radular sac, and a reserve of mature radulae is stored in the short arm, ready for use. The total number of radulae present in a radular sac varies between species and depends on its dietary habits. Thus, molluscivore cones that inject their prey with venom multiple times store a large number of radulae (up to 80). When tactile organs present at the end of the proboscis detect prey, a single-use radula is loaded from the short arm of the radular sac and the prey is harpooned by the radula. The venom is then injected by a powerful contraction of the muscular bulb that releases the contents of the venom gland. Immobilization of the prey is very rapid thanks to the lightning-fast action of the highly neurotoxic venom (on average, a few milliseconds).

Among the 900 or more species of cone described, only the geographer cone (*Conus geographus*) is held responsible for approximately 36 deaths [KOH 16], which has initiated the study of its venom in the 1960s, notably through the work of Alan Kohn in the United States [KOH 60] and above all Robert Endean in Australia [END 65a, END 65b]. Robert Endean demonstrated the initial biological effects of the venom, often very harmful, on the physiology of cones' potential prey (worms, mollusks and fish), as well as on mammals such as mice or guinea pigs [END 74a, END 74b, END 76, END 77a, END 77b, END 79]. He was among the first to point out that the venoms of piscivorous species are more powerful on vertebrates, such as fish and mice, whereas the venoms of molluscivorous species and vermivores produced notable effects mainly on invertebrates. On the basis of his results, Baldomero "Toto" Olivera undertook the systemic characterization of the molecules comprising these venoms, in order to determine the nature and pharmacology of these toxins. In a seminal study, with the use of a chromatography column, he separated the different toxins present in the venom of *Conus geographus* and injected these individually into mice [OLI 90]. The result shows a multitude of behaviors in the rodents, each associated with the type of toxin present in the injected fraction. Some show uncontrollable shaking or convulsions, while others enter into a state of general paralysis, a state of torpor or phases of itching.

Radulae and diet

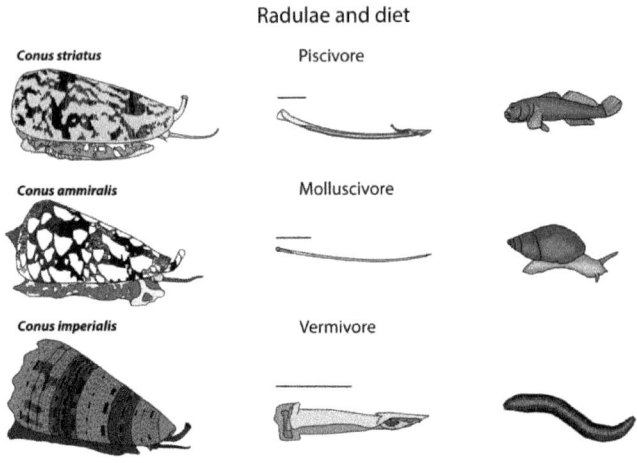

Anatomy of the venomous apparatus of a cone

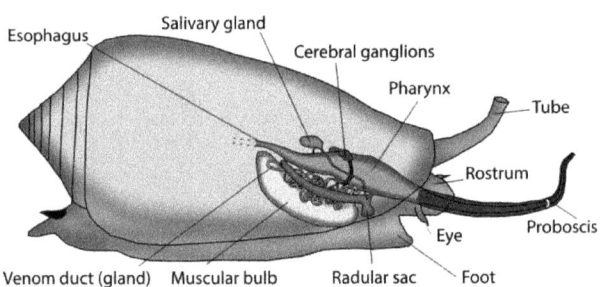

Figure 13.2. *Radula of cones and anatomy of venomous apparatus*

The toxins that make up cone's venoms are called conotoxins. The first of these "conotoxins" to have been characterized, called GI, showed a paralytic action on skeletal muscle and revealed its peptidic nature [GRE 81]. In general, conotoxins are mainly neuro-active peptides rich in cysteine (containing two or more disulfide bridges) that contain between 10 and 35 amino acids, although longer peptides and proteins have also been discovered [LEW 12]. Conotoxins have a wide range of pharmacological targets, including ion channels and neuronal receptors, many of which have been involved in signaling pathways for pain and in the pathophysiology of neurodegenerative diseases such as Parkinson's disease and Alzheimer's disease [LEW 12]. Among the most significant targets of these conotoxins are the acetylcholine nicotinic receptors (α-conotoxins), calcium channels and voltage-dependant sodium channels (ω- and μ-conotoxins), G-protein coupled receptors (contulakins, conopressins), norepinephrine transporters (χ-conopeptide) and N-methyl-D-aspartate receptors (conantokins) [DUT 10a]. Conotoxins are exceptionally potent and capable of discriminating between different subtypes of receptors [LEW 12]. Their small size and the relative ease of synthesis combined with their remarkable pharmacological properties have made conotoxins a precious source of pharmacological tools and potential therapeutics [PRA 12]. In 2004, the FDA approved Prialt™, the first medicine derived from a conotoxin (ω-MVIIA) isolated from the venom of *Conus magus,* which blocks calcium channels [MIL 04]. This medication is proposed as an intrathecal analgesic to treat chronic pain, while other conotoxins are being tested in clinical trials to treat epilepsy and even vascular accidents in the brain [LEW 12, OLI 01]. Despite this progress, conotoxins remain a relatively underexploited source of therapeutic candidates given their remarkable diversity, with estimations suggesting that only 0.1% of the pool of conotoxins has been investigated to date [LEW 12, PRA 144].

The venom of cones is therefore a complex blend of conotoxins, often targeting essential ion channels and whose concerted action (cabal) makes it possible to immobilize prey or deter a predator. Three main cabals of conotoxins have been proposed for a limited number of piscivorous species belonging to one of the three most studied subgenera, namely *Pionoconus*, *Gastridium* and *Chelyconus*. *Pionoconus* and *Chelyconus* use a very original strategy to capture prey, commonly called hook-and-line, involving the extremely rapid injection of a venom containing two types of cabal. The first, called the "lightning-strike cabal", comprises a δ-conotoxin that slows down the inactivation of voltage-dependant sodium channels and a κ-conotoxin that blocks voltage-dependent potassium channels [TER 96]. The concerted activity of these two peptides triggers a series of action potentials in peripheral neurons, causing uncontrollable muscle spasms of an epileptic nature, leading to spastic paralysis. The second cabal, called the "motor cabal", comprises peptides that target the neuromuscular junction, such as ω-conotoxins that block presynaptic voltage-dependant calcium channels (Ca_v channels), α-conotoxins that block postsynaptic acetylcholine nicotinic receptors (nAChR) and μ-conotoxins that block voltage-dependant sodium channels (Na_v channels) expressed at the surface of postsynaptic muscular cells [OLI 01]. The injection of these paralytic conotoxins therefore induces a flaccid paralysis, which follows tetanic immobilization caused by conotoxins from the lightning strike cabal. Surprisingly, two species belonging to the subgenus *Gastridium*, *C. geographus* and *Conus tulipa*, use a strategy called "net strategy", in which they engulf prey that has been sedated by the effects of a venom seemingly secreted directly into the aqueous environment and designed to inhibit the prey's escape response ("nirvana cabal") [OLI 01]. In this cabal, conopeptides such as conantokins are found, which block NMDA receptors, contulakins that are agonists of the neurotensin receptor and conoinsulins, which induce a hypoglycemic state in prey. The release of this cabal into the water has not yet been proven and remains subject to controversy. To date, we know little about the conotoxin cabals present in the venoms of vermivorous and molluscivorous cones, as well as their role in the capture of prey or defense against predators.

13.2. Recent developments in systematics

13.2.1. *Species delimitation*

Unlike morphological traits, which are difficult to apprehend, molecular characters are easily formalized, and the use of variable markers makes it possible to highlight differences between species, even those that have appeared recently. It is therefore not uncommon, in Conoidea, that the use of molecular traits shows that species delimitations based on morphological traits are incorrect [DUD 09, PUI 12]. Furthermore, in the context of integrative taxonomy, where species are considered as hypotheses, validated or rejected according to newly applied criteria, characters or methods, the hypotheses put forward initially on the basis of molecular characters are then reinforced by not only the analysis of different data, such as geographic distribution, bathymetric range or ecological preferences, but also morphologic variability, which can, secondarily, provide additional information [PUI 12]. This approach also makes it possible to provide elements allowing us to propose hypotheses about the evolutionary processes behind these events of speciation. It is also interesting to note that these new hypotheses of species proposed in an integrative context also have an impact in other domains of research. We will illustrate this idea with two examples of cone snails: the *Conus ebraeus* complex and cone snails from Cape Verde.

Conus ebraeus is a species well known by the malacologists; it is widespread in the Pacific and Indian Oceans, and was described in 1758 by Linné. On the basis of a single specimen from the Philippines, Bergh described in 1895 a species that is morphologically very similar, *Conus judaeus*. So similar, in fact, that until recently the majority of taxonomists have doubted its validity and treated the latter as a synonym of *Conus ebraeus*. In 2009, Duda *et al.* [DUD 09] demonstrated through an integrative approach combining molecular markers, morphometric analysis of the shell and comparison of the radulae that *Conus ebraeus* and *Conus judaeus* are indeed different species, although indiscernible from their shells. Puillandre *et al.* [PUI 14] confirmed this hypothesis, showing in addition that they are not even sister species and therefore potentially developed their own toxins. However, a better understanding of this taxon could also have consequences in archeology. In 2016, d'Errico and Backwell [DER 16] argued, by means of morphometric approaches, that the famous cone snail of the "Border Cave" in South Africa, known as the oldest jewel to be found in a human tomb, is not *Conus bairstowi*, as originally suggested, but *Conus ebraeus* (Figure 13.3). Knowing that *Conus bairstowi* and *Conus ebraeus* do not live in the same habitat, this taxonomical re-identification could potentially change our conclusions on the habits of the human populations that gathered the shells. However, how can we be sure that this shell is indeed *Conus ebraeus* and not *Conus judaeus*? These two species do not live in the same habitat either, and even if the presence of *C. judaeus* in South Africa is yet to be confirmed, its presence in the Republic of Seychelles suggests that its zone of distribution could stretch to the southern tip of the African continent. The mystery of the *Conus* of Border Cave is therefore still unsolved.

Figure 13.3. *From left to right:* Conus *of Border Cave,* Conus bairstowi, Conus ebraeus, Conus judaeus

Out of the 5,000 or more species of Conoidea currently considered valid, only 637 are present in the red list of the IUCN (www.iucnredlist.org/), that is, approximately 13%. It is more than the mollusk average (8%) but incomparable to the proportion of vertebrate species, whose status has been evaluated (66%). Of these 637 species of Conoidea, 633 are cones, whose statuses were evaluated in 2011, during a workshop gathering several cone specialists [PET 13]. Despite this significant effort, the red list does not include 102 species described before the workshop, or the 150+ described since, and even less so the 200–300 species of cone that potentially remain to be described.

Of the 633 cones on the red list, 88 are Data Defficient (DD), 478 Least Concerned (LC), 26 Near Threatened (NT), 27 Vulnerable (VU), 11 Endangered (EN) and 3 Critically Endangered (CR). The criteria used to place a species in an IUCN category have been developed for vertebrates, and are often difficult to apply to invertebrates, above all marine invertebrates [REG 15]. Even for cones, the information necessary for their evaluation is rarely available (size of distribution area, population size, etc.), and their status (in particular, DD or LC) seems in certain cases to have been attributed by default (see [HOW 14] for a discussion on this subject for amphibians).

Out of the 14 EN or CR species, 7 are endemic to Cape Verde, and 7 to Senegal. All these species (except one) belong to the subgenus *Lautoconus*, endemic to West Africa and the Mediterranean. These species are for the most part easily recognizable morphologically and are often geographically limited to a single cove. Several articles put forward hypotheses to explain the radiation of cone populations in Cape Verde in particular, but none really tested the limits between species: the molecular data, limited to the COI gene or to full mitochondrial genomes, do not seem to confirm them in any case [CUN 08], or indeed seem to disconfirm some of them [ABA 17]. Peters et al. [PET 16] proposed several measures to protect the cone snails of Cape Verde, a country whose tourism is increasing drastically. Notably, they specify that if "cone snails help support a global industry in the trade for specimen shells and shellcraft", they nonetheless refer to M. Tenorio ("pers. com.") to assert that "although some species are targeted by shell collectors this is not yet believed to have had a major impact on the viability of most". These remarks have not, however, prevented them from advising an export ban on all cone snails from Cape Verde, other than for scientific purposes. This will eventually allow scientists to show that the diversity of cone snails from Cape Verde is overestimated, thus diminishing the risks on each species, and rendering the proposed conservation measures, to a certain extent, unsuitable. Yet again, taxonomical hypotheses in Conoidea, often questioned, do not affect taxonomists only.

13.2.2. *Phylogeny and classification*

The expression "molecular revolution" is particularly adapted when talking about Conoidea: rarely a taxon and its phylogenetic relationships have been modified to the extent we have seen in the case of Conoidea because of the contribution of DNA data. The splitting up of the Turridae initiated in 1993 [TAY 93] was confirmed in a classification based on molecular phylogeny, as Conidae were confirmed in their previous acceptation while Turridae were separated into 12 families [BOU 11] (Figure 13.4(a)). Another family has since been described: Bouchetispiridae [KAN 12]. Others will be in the future, in a new classification derived from as yet unpublished works, based on the sequencing of entire mitogenomes and an approach based on exon capture [BI 12].

The methods applied to Conoidea in general have also been used for certain families. Several molecular phylogenies of the Terebridae have been published [CAS 12, MOD 14]: in addition to a revision of the groups within the family, these have made it possible to re-evaluate the evolution of the morpho-anatomical characters traditionally used. Thus, while Taylor [TAY 90] has described several species that have lost their venom gland and their radula and others that have developed a hypodermic radula, similar to the radulae found in cones but different from the other Terebridae, these molecular phylogenies have shown that these events were not unique cases. In fact, the venomous gland may have been lost at least six times, independently, over the course of evolution, and the hypodermic radula may have appeared three times independently (Figure 13.4(b)). This is ample justification to not only question the taxa based on these characters but also overturn hypotheses about the evolution of Terebridae: a group in which characters were supposed to vary very little, with very rare character state changes, is actually capable of profound, repeated modifications, in a relatively short period of time.

Together with Terebridae, several phylogenies of Conidae, based mainly on mitochondrial markers, have been published [PUI 14]. In the case of Conidae, the mitochondrial genomes recently sequenced have made it possible to publish a new phylogeny of the Conidae, with six main lineages, placed at the genus level, the phylogenetic relationships of which being resolved and strongly supported [URI 17]. As with the Terebridae, these phylogenies allow us to propose hypotheses about their evolution. As an example, it has been possible to partially infer the evolution of diet. Diets are known for only three of the six genera: the only species of *Californiconus* feeds on worms, mollusks, crustaceans and fish; species from the genus *Conasprella* feed on worms, but two lineages each changed independently for a piscivorous diet and a molluscivorous diet (Figure 13.4(c)).

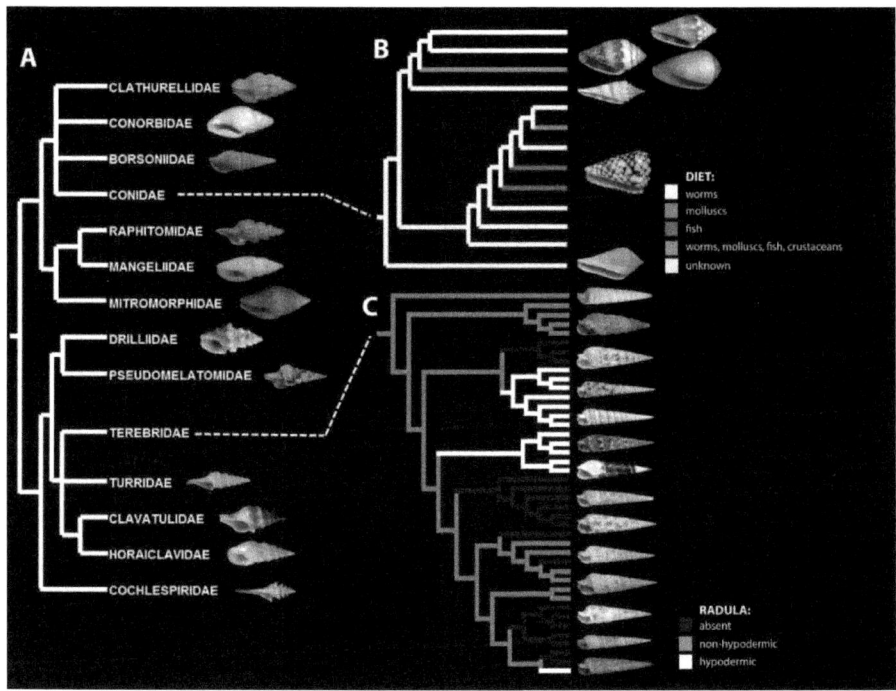

Figure 13.4. *a) Current phylogeny and classification of Conoidea [PUI 11]; b) Phylogeny and evolution of diet in Conidae [PUI 14, URI 17]; c) Phylogeny and evolution of the morphology of radula in Terebridae [CAS 12]. For a color version of this figure, see www.iste.co.uk/grandcolas/biodiversity.zip*

13.3. Toxins: genomic, transcriptomic and proteomic approaches

Until recently, conotoxins have been isolated and characterized mainly by a process called "bioactivity-guided fractionation", wherein venoms are screened against various biological "targets" and the peptides responsible for the detected activity ("hits") are isolated by a series of purification steps and intermediary tests for bioactivity [PRA 12, PRA 14]. This approach requires a disproportionate amount of venom samples to characterize, in many cases, one conotoxin. Not only does this method depend on resources of venom, but it is also tedious and does not allow for holistic studies of the venom's composition [PRA 12, PRA 14]. These inconveniences make these traditional methods, mainly reliant on bioactivity, largely obsolete, in particular when one tries to extract ecological or evolutionary data based on the composition of venoms. Recent technological processes, notably in terms of high-throughput sequencing, as well as the development of advanced techniques of mass spectrometry with higher sensitivity and new bio-informatic tools allowing us to combine and analyze large volumes of data, have progressively supplanted traditional methods [PRA 12]. This integrated method is referred to as a "venomic" approach.

13.3.1. *Methodological developments*

One of the original contributions of proteomics has been to reveal the conotoxins expressed by the different species of cone snails. The first studies on the venoms of *Conus geographus*, *Conus tulipa*, *Conus textile* and *Conus striatus*, which used mass spectrometry coupled with a separation system by upstream HPLC, lead to the first estimation of 100–200 conotoxins per venom [BIN 96]. However, rapid improvements in the separation capacity and sensitivity of the instruments used in mass spectrometry have allowed researchers to observe conotoxins expressed at a very low level for the first time. In fact, very recently, the use of this "venomic" approach applied to the venom of *C. marmoreus* has revealed the presence of more than 8,000 peptides in this species and made it possible to highlight the process behind this enormous diversity [DUT 13]. This era of venomics has revealed variations not only in the composition of venom between different cone species but also within the same species, often collected from the same location. The intra-specific variations have been demonstrated in *Conus ventricosus, C. textile, C. vexillum, C. marmoreus, C. consors* and *C. imperialis* [DAV 09, DUT 10a, ROM 08].

Although proteomic approaches have become a keystone in the study of conotoxin diversity, approaches based on the cloning of nucleic sequences have been applied to the discovery of new conotoxins as well as to understand the evolutionary mechanisms of cone venoms. The approach of a cDNA library was initially used to reveal the architecture of the conotoxins' standard precursor. It was then used to define the notion of a gene superfamily based on the conservation of the sequence signal as well as to reconstruct the evolutionary relationships of different clades. However, this method has several limitations. For example, the scope is limited to targeted sequences and, as a result, the rarer or new transcripts are often ignored; the method is thus incapable of producing sequencing data that cover the entire composition of the venom.

To overcome these limitations, high-throughput sequencing technologies have been used to sequence the complete transcriptomes of cones' venom glands. Such sequencing technologies are also used to sequence entire genomes. Roche's 454 pyrosequencing platform offered longer sequences (up to 700pb), while Illumina platforms widely used today produce shorter sequences but with a much higher coverage. These two platforms have been used in tandem to sequence the transcriptome and genome of *Conus bullatus* [HU 11], while the venom gland transcriptome of *Conus consors* was sequenced using only 454 pyrosequencing [TER 12].

Although only a limited number of transcriptomes from cone species have been sequenced to date, this number should increase drastically in the future. The quantity of data produced by high-throughput sequencing remains a major challenge for researchers in this area. However, the extraction of conotoxin sequences from sequencing data has been facilitated by the elaboration of specifically designed bioinformatic tools. These programs, based on the research of sequence similarities using BLAST and Markov models (pHMM) such as ConoDictor and ConoSorter, were published to help accelerate the process of discovering more new conotoxins as well as new superfamilies of conotoxins [KAA 12, KOU 12, LAV 13]. For example, revising the set of data on the transcriptome of *C. marmoreus* using ConoSorter made it possible to identify 158 new sequences of conotoxins belonging to 13 superfamilies of genes [LAV 13]. These recent progresses have led to a veritable deluge of sequences, among which some show new cysteine patterns probably forming new structural motifs, and which could interact with yet unknown physiological targets.

13.3.2. *Recent discoveries*

Historically, researchers first looked into the venoms of cone snails to understand their lethal effect on humans. In fact, numerous medical reports pointed out the defensive behavior of cones, which has led to various incidents with humans, including approximately 30 recorded deaths [KOH 16]. However, among the various species of cone, only stings from *C. geographus* have caused death, and to date, there is still no antidote [DUT 14a]. The secrets of this aggressive species' potent venom have been partially discovered thanks to the use of venom gland extracts; these extracts showed highly paralytic components [END 74b]. The three major classes of paralytic toxins targeting neuromuscular junctions have been isolated, namely α-conotoxins, μ-conotoxins and ω-conotoxins, which block, respectively, acetylcholine nicotinic receptors and voltage-dependent sodium and calcium channels. This group of conotoxins makes up the "motor cabal" whose role in capturing prey is widely documented. An even more surprising fact is that by separately gathering the predation and defense injected venoms of *C. geographus*, it has been recently demonstrated that these fatal, paralytic toxins were only abundant in defensive stings and absent in venom that is designed to be injected into prey. The latter contained, on the contrary, toxins specific to the physiology of its prey (fish), which show a weak activity on mammalian targets.

Another recent discovery has revealed the presence of peptides resembling insulin in the venom of the geographer cone [SAF 15, SAF 16]. These conoinsulins may have evolved to imitate the insulin of their prey rather than endogenous mollusk insulin. Another class of conopeptides widely spread in this species are conantokins, antagonists of NMDA receptors. Along with other non-paralytic conopeptides, including agonists of vasopressin receptors, conopressins, they seem to play a role in the prey capture strategy developed by *C. geographus*. The hypothesis has been put forward that these peptides are secreted in the immediate surroundings of fish to sedate them and then capture them in their "net". However, the secretions of these species have not yet been collected and analyzed to demonstrate potential synergies between these different conopeptides. Thus, more fully developed in-depth studies considering recent ecological and behavioral discoveries are necessary to fully understand the synergic combinations between different conotoxins.

13.4. Evolution of Conoidea: integrative approaches

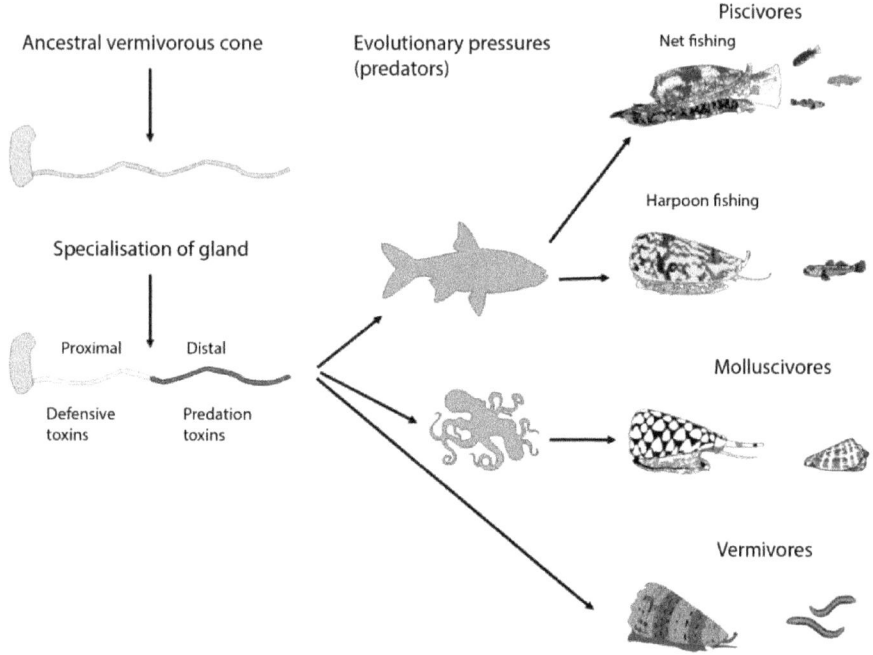

Figure 13.5. *Vermivorous/piscivorous evolutionary model*

The evolution of a species is dictated in part by the sum of its biotic interactions, with predatory, competitive and defensive reactions among the most frequent and influential ones. Cones, in addition to using their venoms mainly to capture prey and their shell to provide protection against predators, may, only in situations of extreme necessity, use their venoms defensively [DUT 14b, JIN 15, PRA 16]. As previously described, venoms used for one or the other function are usually complex mixtures of conotoxins, often targeting essential ion channels and which have been supposed to work in synergy (cabals). The pharmacology of the "lighting cabal" has served as a basis for the hypothesis of the evolution of piscivorous cones [TER 96]. δ-conotoxins, which slow down the inactivation of Na_v channels, are not only widely spread in species that feed on fish, but have also been identified in species that feed on mollusks and worms. In light of the knowledge that the ancestor of cones was probably a vermivore, as the phylogenic approach shows, the hypothesis that δ-conotoxins were first used to deter fish from competing with ancestral vermivorous cones for the same prey (worms) was initially proposed [AMA 15]. The simultaneous injection of δ-conotoxins with a subsequent κ-conotoxin would cause spastic paralysis when injected in the fish, which would allow for the predation of fish. It has been suggested that different subgenera of piscivorous cones developed different classes of κ-conotoxins in a convergent fashion. For example, *C. purpurascens* from the subgenus *Chelyconus* uses a peptide belonging to the superfamily O1 called PVIIA, which contains an ICK motif, while *Conus striatus* (*Pionoconus*) uses a peptide containing the kunitz domain called Conkunitzin-S1, to the same end. A third type, κM-RIIIK from *Conus radiatus* belonging to the M superfamily, which is usually associated with the blockage of Na_v channels, has also been identified.

However, the ability of certain cone species to inject different venoms for predation and defense, as well as the role of defense in the evolution of conotoxins, has led to a re-examination of this hypothesis [JIN 15]. In fact, the venom of the species of each group (piscivores, molluscivores and vermivores) can be obtained after a defense stimulus, which would suggest that this adaptation is relatively ancestral [DUT 14b]. The injection of different venoms is facilitated by the compartmentalized venom gland, an original anatomical feature that also seems to be an ancestral feature. Given that paralytic conotoxins from the "motor" cabal, which have been supposed to be used in the capture of prey by piscivorous cones like *C. geographus*, have in fact been found to be injected for the most part in "defense" venom, a new hypothesis has emerged, wherein "defensive" toxins have provided a molecular basis allowing for this remarkable change of prey evident in the phylogeny of cones, from slow-moving invertebrates to much faster vertebrates. These different hypotheses require a re-interpretation of the evolutionary history of cones, particularly through a pharmacological "lens", where major conotoxin families will be systematically evaluated according to their ethological and ecological roles in the defense and/or predation of these animals.

This is just one example of the numerous evolutionary hypotheses proposed to explain the diversification of cones and Conoidea (see on this subject numerous articles co-written by T. Duda Jr.). However, in all cases, we have been able to put forward these hypotheses and, if only in part for the moment, to test them thanks to the combination of approaches from systematics, which not only provide taxonomic and phylogenetic hypotheses, but also allow us to understand the evolution of the toxinological, physiological, ecological, anatomic and even behavioral characteristics of these interesting organisms.

13.5. Bibliography

[ABA 17] ABALDE S., TENORIO M.J., AFONSO C.M.L. *et al.*, "Mitogenomic phylogeny of cone snails endemic to Senegal", *Molecular Phylogenetics and Evolution*, vol. 112, pp. 79–87, 2017.

[AMA 15] AMAN J.W., IMPERIAL J.S., UEBERHEIDE B. *et al.*, "Insights into the origins of fish hunting in venomous cone snails from studies of *Conus tessulatus*", *Proceedings of the National Academy of Sciences of the United States of America*, vol. 112, pp. 5087–5092, 2015.

[BAC 12] BACHER S., "Still not enough taxonomists: reply to Joppa *et al.*", *Animal Conservation*, vol. 5, pp. 245–249, 2012.

[BI 12] BI K., VANDERPOOL D., SINGHAL S. *et al.*, "Transcriptome-based exon capture enables highly cost-effective comparative genomic data collection at moderate evolutionary scales", *BMC Genomics*, vol. 13, p. 403, 2012.

[BIN 96] BINGHAM J.P., JONES A., LEWIS R.J. *et al.*, Conus *venom peptides (conopeptides): inter-species, intra-species and within individual variation revealed by ionspray mass spectrometry*, in LAZAROVICI E., SPIRA M.E., ZLOTKIN E. (eds), "Biomedical Aspects of Marine Pharmacology", pp. 13–27, Alaken, Fort Collins, 1996.

[BOU 11] BOUCHET P., KANTOR Y., SYSOEV A. *et al.*, "A new operational classification of the Conoidea (Gastropoda)", *Journal of Molluscan Studies*, vol. 77, pp. 273–308, 2011.

[CAS 12] CASTELIN M., PUILLANDRE N., KANTOR Y. *et al.*, "Macroevolution of venom apparatus innovations in auger snails (Gastropoda; Conoidea; Terebridae)", *Molecular Phylogenetics and Evolution*, vol. 64, pp. 21–44, 2012.

[CUN 08] CUNHA R.L., TENORIO M.J., AFONSO C. *et al.*, "Replaying the tape: recurring biogeographical patterns in Cape Verde *Conus* after 12 million years", *Molecular Ecology*, vol. 17, pp. 885–901, 2008.

[DAV 09] DAVIS J., JONES A., LEWIS R.J., "Remarkable inter- and intra-species complexity of conotoxins revealed by LC/MS", *Peptides*, vol. 30, pp. 1222–1227, 2009.

[DER 16] D'ERRICO F., BACKWELL L., "Earliest evidence of personal ornaments associated with burial: the *Conus* shells from Border Cave", *Journal of Human Evolution*, vol. 93, pp. 91–108, 2016.

[DUD 09] DUDA T.F. JR., KOHN A.J., MATHENY A.M. et al., "Cryptic species differentiated in Conus ebraeus, a widespread tropical marine gastropod", Biological Bulletin, vol. 217, pp. 292–305, 2009.

[DUT 10a] DUTERTRE S., BIASS D., STOCKLIN R. et al., "Dramatic intraspecimen variations within the injected venom of Conus consors: an unsuspected contribution to venom diversity", Toxicon, vol. 55, pp. 1453–1462, 2010.

[DUT 10b] DUTERTRE S., LEWIS R.J., "Use of venom peptides to probe ion channel structure and function", The Journal of Biological Chemistry, vol. 285, pp. 13315–13320, 2010.

[DUT 13] DUTERTRE S., JIN A.H., KAAS Q. et al., "Deep venomics reveals the mechanism for expanded peptide diversity in cone snail venom", Molecular & Cellular Proteomics, vol. 12, pp. 312–329, 2013.

[DUT 14a] DUTERTRE S., JIN A.H., ALEWOOD P.F. et al., "Intraspecific variations in Conus geographus defence-evoked venom and estimation of the human lethal dose", Toxicon, vol. 91, pp. 135–144, 2014.

[DUT 14b] DUTERTRE S., JIN A.H., VETTER I. et al., "Evolution of separate predation- and defence-evoked venoms in carnivorous cone snails", Nature Communications, vol. 5, p. 3521, 2014.

[END 65a] ENDEAN R., IZATT J., "Pharmacological study of the venom of the gastropod Conus magus", Toxicon, vol. 3, pp. 81–93, 1965.

[END 65b] ENDEAN R., RUDKIN C., "Further studies of the venoms of conidae", Toxicon, vol. 2, pp. 225–249, 1965.

[END 74a] ENDEAN R., GYR P., PARISH G., "Pharmacology of the venom of the gastropod Conus magus", Toxicon, vol. 12, pp. 117–129, 1974.

[END 74b] ENDEAN R., PARISH G., GYR P., "Pharmacology of the venom of Conus geographus", Toxicon, vol. 12, pp. 131–138, 1974.

[END 76] ENDEAN R., WILLIAMS H., GYR P. et al., "Some effects on muscle and nerve of crude venom from the gastropod Conus striatus", Toxicon, vol. 14, pp. 267–274, 1976.

[END 77a] ENDEAN R., GYR P., SURRIDGE J., "The pharmacological actions on guinea-pig ileum of crude venoms from the marine gastropods Conus striatus and Conus magus", Toxicon, vol. 15, pp. 327–337, 1977.

[END 77b] ENDEAN R., SURRIDGE J., GYR P., "Some effects of crude venom from the cones Conus striatus and Conus magus on isolated guinea-pig atria", Toxicon, vol. 15, pp. 369–374, 1977.

[END 79] ENDEAN R., GYR P., SURRIDGE J., "The effects of crude venoms of *Conus magus* and *Conus striatus* on the contractile response and electrical activity of guinea-pig cardiac musculature", *Toxicon*, vol. 17, pp. 381–395, 1979.

[GRA 81] GRAY W.R., LUQUE A., OLIVERA B.M. *et al.*, "Peptide toxins from *Conus geographus* venom", *The Journal of Biological Chemistry*, vol. 256, pp. 4734–4740, 1981.

[HOW 14] HOWARD S.D., BICKFORD D.P., "Amphibians over the edge: silent extinction risk of Data Deficient species", *Diversity and Distributions*, vol. 20, pp. 837–846, 2014.

[HU 11] HU H., BANDYOPADHYAY P.K., OLIVERA B.M. *et al.*, "Characterization of the *Conus bullatus* genome and its venom-duct transcriptome", *BMC Genomics*, vol. 12, p. 60, 2011.

[JIN 15a] JIN A.H., ISRAEL M.R., INSERRA M.C. *et al.*, "Delta-conotoxin SuVIA suggests an evolutionary link between ancestral predator defence and the origin of fish-hunting behaviour in carnivorous cone snails", *Proceedings Biological Sciences of the The Royal Society*, vol. 282, p. 20150817, 2015.

[JIN 15b] JIN A.H., VETTER I., HIMAYA S.W. *et al.*, "Transcriptome and proteome of *Conus planorbis* identify the nicotinic receptors as primary target for the defensive venom", *Proteomics*, vol. 15, pp. 4030–4040, 2015.

[KAA 12] KAAS Q., YU R., JIN A.H. *et al.*, "ConoServer: updated content, knowledge, and discovery tools in the conopeptide database", *Nucleic Acids Research*, vol. 40, pp. D325-D330, 2012.

[KAN 12] KANTOR Y.I., STRONG E.E., PUILLANDRE N., "A new lineage of Conoidea (Gastropoda: Neogastropoda) revealed by morphological and molecular data", *Journal of Molluscan Studies*, vol. 78, pp. 246–255, 2012.

[KOH 60] KOHN A.J., SAUNDERS P.R., WIENER S., "Preliminary studies on the venom of the marine snail *Conus*", *Annals of the New York Academy of Sciences*, vol. 90, pp. 706–725, 1960.

[KOH 16] KOHN A.J., "Human injuries and fatalities due to venomous marine snails of the family Conidae", *International Journal of Clinical Pharmacology and Therapeutics*, vol. 54, pp. 524–538, 2016.

[KOU 12] KOUA D., BRAUER A., LAHT S. *et al.*, "ConoDictor: a tool for prediction of conopeptide superfamilies", *Nucleic Acids Research*, vol. 40, pp. W238–W241, 2012.

[LAV 13] LAVERGNE V., DUTERTRE S., JIN A.H. *et al.*, "Systematic interrogation of the *Conus marmoreus* venom duct transcriptome with ConoSorter reveals 158 novel conotoxins and 13 new gene superfamilies", *BMC Genomics*, vol. 14, p. 708, 2013.

[LEW 12] LEWIS R.J., DUTERTRE S., VETTER I. *et al.*, "*Conus* venom peptide pharmacology", *Pharmacological Reviews*, vol. 64, pp. 259–298, 2012.

[MIL 04] MILJANICH G.P., "Ziconotide: neuronal calcium channel blocker for treating severe chronic pain", *Current Medicinal Chemistry*, vol. 11, pp. 3029–3040, 2004.

[MOD 14] MODICA M.V., PUILLANDRE N., CASTELIN M. *et al.*, "A good compromise: rapid and robust species proxies for inventorying biodiversity hotspots using the terebridae (Gastropoda: Conoidea)", *PLoS ONE*, vol. 9, p. e102160, 2014.

[OLI 90] OLIVERA B.M., RIVIER J., CLARK C. *et al.*, "Diversity of *Conus* neuropeptides", *Science*, vol. 249, pp. 257–263, 1990.

[OLI 01] OLIVERA B.M., CRUZ L.J., "Conotoxins, in retrospect", *Toxicon*, vol. 39, pp. 7–14, 2001.

[PET 13] PETERS H., O'LEARY B.C., HAWKINS J.P. *et al.*, "*Conus*: first comprehensive conservation red list assessment of a marine gastropod mollusc genus", *PLoS ONE*, vol. 8, p. e83353, 2013.

[PET 16] PETERS H., O'LEARY B.C., HAWKINS J.P. *et al.*, "The cone snails of Cape Verde: marine endemism at a terrestrial scale", *Global Ecology and Conservation*, vol. 7, pp. 201–213, 2016.

[POW 66] POWELL A.W.B., "The molluscan families Speightiidae and Turridae. An evaluation of the valid taxa, both Recent and fossil, with lists of characteristics species", *Bulletin of the Auckland Institute and Museum*, vol. 5, pp. 5–184, 1966.

[PRA 12] PRASHANTH J.R., LEWIS R.J., DUTERTRE S., "Towards an integrated venomics approach for accelerated conopeptide discovery", *Toxicon*, vol. 60, pp. 470–477, 2012.

[PRA 14] PRASHANTH J.R., BRUST A., JIN A.H. *et al.*, "Cone snail venomics: from novel biology to novel therapeutics", *Future Medicinal Chemistry*, vol. 6, pp. 1659–1675, 2014.

[PRA 16] PRASHANTH J.R., DUTERTRE S., JIN A.H. *et al.*, "The role of defensive ecological interactions in the evolution of conotoxins", *Molecular Ecology*, vol. 25, pp. 598–615, 2016.

[PUI 11] PUILLANDRE N., KANTOR Y., SYSOEV A. *et al.*, "The dragon tamed? A molecular phylogeny of the Conoidea (Mollusca, Gastropoda)", *Journal of Molluscan Studies*, vol. 77, pp. 259–272, 2011.

[PUI 12] PUILLANDRE N., MODICA M.V., ZHANG Y. *et al.*, "Large scale species delimitation method for hyperdiverse groups", *Molecular Ecology*, vol. 21, pp. 2671–2691, 2012.

[PUI 14] PUILLANDRE N., BOUCHET P., DUDA T.F. *et al.*, "Molecular phylogeny and evolution of the cone snails (Gastropoda, Conoidea)", *Molecular Phylogenetics and Evolution*, vol. 78, pp. 290–303, 2014.

[RÉG 15] RÉGNIER C., ACHAZ G., LAMBERT A. *et al.*, "Mass extinction in poorly known taxa", *Proceedings of the National Academy of Sciences*, vol. 112, p. 7761–7766, 2015.

[ROM 08] ROMEO C., DI FRANCESCO L., OLIVERIO M. *et al.*, "*Conus ventricosus* venom peptides profiling by HPLC-MS: a new insight in the intraspecific variation", *Journal of Separation Science*, vol. 31, p. 488–498, 2008.

[SAF 15] SAFAVI-HEMAMI H., GAJEWIAK J., KARANTH S. *et al.*, "Specialized insulin is used for chemical warfare by fish-hunting cone snails", *Proceedings of the National Academy of Sciences of the United States of America*, vol. 112, pp. 1743-1748, 2015.

[SAF 16] SAFAVI-HEMAMI H., LU A., LI Q. *et al.*, "Venom insulins of cone snails diversify rapidly and track prey taxa", *Molecular Biology and Evolution*, vol. 33, pp. 2924-2934, 2016.

[TAY 90] TAYLOR J.D., "The anatomy of the foregut and relationships in the Terebridae", *Malacologia*, vol. 32, pp. 19–34, 1990.

[TAY 93] TAYLOR J.D., KANTOR Y.I., SYSOEV A.V., "Foregut anatomy, feedings mechanisms and classification of the Conoidea (= Toxoglossa) (Gastropoda)", *Bulletin of the Natural History Museum of London (Zoology)*, vol. 59, pp. 125–170, 1993.

[TER 96] TERLAU H., SHON K.J., GRILLEY M. *et al.*, "Strategy for rapid immobilization of prey by a fish-hunting marine snail", *Nature*, vol. 381, pp. 148–151, 1996.

[TER 12] TERRAT Y., BIASS D., DUTERTRE S. *et al.*, "High-resolution picture of a venom gland transcriptome: case study with the marine snail *Conus consors*", *Toxicon*, vol. 59, pp. 34–46, 2012.

[TRY 84] TRYON G.W., *Manual of Conchology; Structural and systematic, with illustrations of the species. Vol. VI. Conidae, Pleurotomidae*, Tryon G.W., Philadelphia, 1884.

[URI 17] URIBE J.E., PUILLANDRE N., ZARDOYA R. *et al.*, "Beyond *Conus*: Phylogenetic relationships of conidae based on complete mitochondrial genomes", *Molecular Phylogenetics and Evolution*, vol. 107, pp. 142–151, 2017.

The Anthropocene: a Geological or Societal Subject?

Geology is a discipline in which time plays a fundamental role. Indeed, it forms the very backbone of this discipline. Furthermore, as it covers a timespan of over 4 billion years, it is necessary to divide this time into separate intervals. The only record of Earth's history, its past time, is the rock record. And this rock record is subdivided into stratigraphic intervals of varying thickness and referred to by a hierarchal set of units (Erathems, Systems, Series and Stages). These material rock units (i.e. stratigraphic intervals) serve as the basis for the corresponding units of the Geologic Time Scale (Eras, Periods, Epochs, and Ages). These stratigraphic intervals are more or less thick, and they serve to establish correlations between different geographical terrains that are often very distant from one another. Since life spread over the Earth in abundance, 550 million years ago (Ma), Erathems, Systems, Series, and Stages are distinguished on the basis of their fossil content, and more generally by crises in the living world.

Thus, the Paleozoic Era is separated from the Mesozoic era by the biggest crisis the living world has ever known (in which 96% of species disappeared) and certain iconic organisms disappeared forever, such as tribolites or certain groups of marine unicellular organisms like the Fusulinacea. The Mesozoic Era is also separated from the Cenozoic Era by another major biological crisis that saw the disappearance of all non-avian dinosaurs, ammonites, etc.

Chapter written by Patrick DE WEVER and Stan FINNEY.

All these crises the living world suffered from are associated with profound disturbances in the environment (variation in sea level of 100–200 m, change in the atmosphere's CO_2 and SO_2 contents, global cooling or warming, etc.). Smaller subdivisions of the geological timescale (Ages) are based on identical principles, but of less significance.

The establishment of the different subdivisions and their definition are thus in accordance with very precise norms and methods, managed and guided by commissions from the IUGS (International Union of Geological Sciences), one of the objectives of which is to establish *standards* so that the global community uses words that have the same meaning everywhere. The subdivisions of geological time are part of these standards. The proposal for each new subdivision is studied by a task group that submits it to a sub-commission, before it is submitted to a commission and finally to an executive comity before being ratified and finally introduced into the geological timescale (Figure 14.1). The procedure is long and followed with the same rigor as that used by States to change a law.

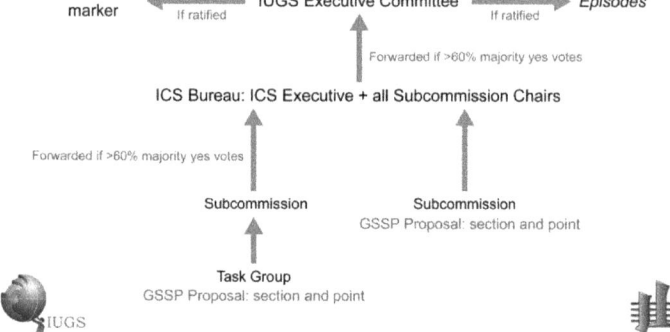

Figure 14.1. *Procedure followed to modify a name or an age or to introduce/remove a subdivision, from the initial examination by a task group. Regarding the question of the Anthropocene, the procedure should be: proposition of the task group (current), submission to the sub-commission of the Quaternary (not yet carried out), commission of stratigraphy, etc.*

Let us take as an example the Quaternary. This subdivision of the geological timescale, introduced in 1829 by Jules Desnoyers, took several decades to truly establish itself (the British geologist Lyell never accepted it). In 1982, after a decade of discussions between specialists, the sub-commission of the Quaternary decided to officially pose the question of its lower limit at a congress in Moscow. In 1995, the American William Berggren suggested abandoning the term, a proposition adopted in 2004; it was no longer figured in the international geological timescale at the International Geological Congress in Florence. After 4 years, the subject was nonetheless taken up again and the word Quaternary was reintroduced into the international geological timescale, but as a system this time, not an era. It figured again on the timescale at the next geological congress at Oslo in 2008. However, its case was not yet fixed, as then its lower limit was still subject to debate. It could have been either 2,588 million years or 1,806 million years, according to whether the Gelasian stage is a part of it or not. The final decision was made in 2009 [GIB 09]. It took 27 years, from 1982 to 2009, for the community to agree upon this international standard.

A dossier concerning precise criteria could then be presented to the commission ad hoc. Among these criteria, some are geological (sedimentary continuity, rate of sedimentation, etc.), others are biostratigraphic (high modification of fauna during a significant length of time in the case of abundant and diverse fossils, etc.) and others have to do with chemical physics (isotopes, magnetics, etc.). However, especially, criteria that are global and synchronous at the same time are necessary. Synchronism is key, as the distinctions have been used, first and foremost, to date.

14.1. A new "geological era"?

The word Anthropocene was coined at the beginning of the 20th Century, but it was popularized in 1995 by Paul Crutzen (Nobel laureat in chemistry) to designate a period during which anthropic activities left an impression on the entire planet (as we will see later). This term has since flourished in scientific literature and, perhaps to an even greater extent, in social and political sciences and, especially, in the media. Google Scholar, for example, gives us (as of 27 April 2017) 40,100 articles for a notion introduced in 2002 (it increases by 50 every day)! The success of this term lies in the fact that it is widely copied. Thus, for example, in a book entitled "The Anthropocene Event" written by two historians, Christophe Bonneuil and Jean-Baptiste Fressoz [BON 13], the authors do not hesitate to name it under the Thermocene, the Thanatocene, the Phagocene, the Phronocene and the

Polemocene. Since the beginning of 2016, the political media have made every attempt to not be left out of this phenomenon. It is in such a climate that the newspaper The Guardian had talked about the "Trumpocene". The newspaper Le Monde also addressed this after the unexpected election of the president of the United States in its 14 November 2016 edition.

Some people would like to make this a geological era because the influence of humankind is global. For this reason, the name is constructed the same way as other subdivisions of the geological timescale, such as the "series" entitled: the Eocene (−56 to 48 Ma); the Oligocene (−34 to −23 Ma); the Miocene (−23 to −5.3 Ma); the Pliocene (−5.3 to −2.5 Ma); the Pleistocene (−2.6 to 0.012 Ma) and the Holocene (since 0.012 Ma).

The Anthropocene would therefore follow on from the Holocene and would only be 0.00005 Ma old, a different order of magnitude. Furthermore, calling it a "geological" subdivision would seem to highlight the importance of its influence; however, this also entails forgetting certain fundamental elements. In order to be accepted, a subdivision of the geological timescale must respect a certain number of precise criteria before the dossier is submitted to examination for eventual ratification (Figure 14.1).

In addition, it seems strange to talk about it as an "era", one of the main subdivisions in the geological timescale. In fact, the shortest geological era is 65 million years long. In the case of the Anthropocene, we are not at all on the same scale in terms of time duration. Does the appearance of a term subsequently authorize a misleading amalgam of time? The confusions are doubtlessly due to the fact that Paul Crutzen, who is not a geologist, used this word in the vernacular sense, as one would say the Christian, industrial or atomic era. For geologists, an era is tens of millions of years long.

In geology, as in biology or history, the positioning of limits is a delicate art. Recently, stratigraphers have been in conflict to the way in which we have stratigraphically divided glacial oscillations for the past 2.8 million years. Those who wanted to add a subdivision into this part of human history prided themselves on the recognition granted them in 2006, when the Geological Society of London[1] asked the question "Do we now live in the Anthropocene?" The 21 members of its commission answered positively, invoking the fact that the Holocene is over and that the Earth has entered "a stratigraphic interval without comparable precedent over the course of the

---

1 It tends to call itself the Geological Society no doubt to appear to be The Geological Society.

last few millions years". Together with the accumulation of greenhouse gases, they evoke the human transformation of countryside, which "now noticeably surpasses natural sedimentary production", as well as the acidification of oceans and the destruction of wildlife. This decision is not at all legitimate, since we have seen that to be recognized a unit of the geological time scale it must be validated by an international commission[2].

The subjects of her Majesty are opinionated people and, since then, have multiplied their articles to this effect. Above all, they inundate the media with messages, having well understood that communication prevails over reason and/or reflection (they did not need to wait for D. Trump to understand this). Thus, since 2012, during the time of the Geological Congress held in Brisbane, we have seen journalists and scientists write that this point was going to be debated and voted on. Rumors came solely from the media, as this discussion was not on the agenda! The dossier was not scientifically informed. The president of the International Stratigraphic Commission (co-author of the present article) had specified that it would perhaps not even be ready before the next congress, in 2016. He was also surprised that this proposition was put forward by scientists who do not in fact all have a clear idea of what a geological subdivision represents and what one requires (in terms of landmarks, recording findings in sediments, etc.).

14.2. Criteria to distinguish the Anthropocene

For those who support the notion of the Anthropocene, this period starts with the first impressions left by humankind. This was first generally accepted to be the middle of the 19th Century, with the Industrial Revolution. Yet, some scientists wanted to refine their analysis to find a more precise date, and proposals then appeared in abundance.

Some, following on from the chemist Paul Crutzen, have this epoch start in 1784, the year of the patent of James Watt's[3] steam engine that laid the

---

2 The British had a large Empire, might they have conserved some of the habits that come with one?

3 The steam engine as improved and patented by James Watt, for example, results from a long process of evolution between 1765 and 1780 that made it possible to move on from a machine of limited use, in the middle of the century, to an efficient machine with numerous applications toward the end of the same century. It was the source of mechanical energy for the nascent Industrial Revolution. It was essential for water pumping in mines, for bringing up coal and then in the advances that have followed in the area of transport, such as the steamboat or the steam locomotive.

foundations for the Industrial Revolution. However, others proposed different beginnings: the dawn of the 20th Century, the Renaissance[4] or even the Neolithic. The phenomena invoked to delimit the Anthropocene are recent on a geological scale, but include landmarks covered in signs of a longstanding diachrony in terms of modification of the countryside and the biotope by human activity since Antiquity (Figures 14.2 and 14.3).

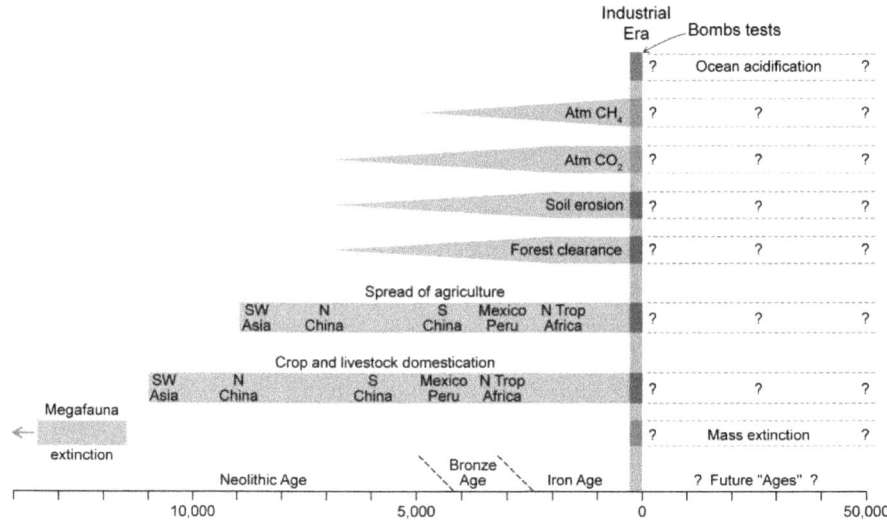

Figure 14.2. *Human impacts on the environment at the Earth's surface (according to [RUD 15]). The impacts are not synchronous for each of the categories, neither are they within each category as regards the places concerned (e.g. domestication began to manifest itself 11,000 years ago in Southwest Asia but only 3,000 years ago in Tropical Africa). For a color version of this figure, see www.iste.co.uk/grandcolas/biodiversity.zip*

4 According to either the authors or the facts taken into account, this period is of varying length. For some, the Renaissance starts with Petrarch (1303–1374), for others it starts in 1415, with the first Portuguese implantation in North Africa, in the 1450s with the invention of the printing press by Gutenberg, in 1453 with the fall of Constantinople (a date retained in French academia to designate the end of the Middle Ages and the start of the Renaissance) or in 1492 with the capture of Grenada (end of the Spanish Reconquista when Isabel I of Castile eliminated the last Muslim kingdom from the Spanish peninsula, at the same time as the discovery of America by Christopher Columbus). The dates of its end are just as variable: the death of Charles V (1558), that of Giordano Bruno (1600), that of Henry IV (1610), a date retained by the French, that of the English playwright Shakespeare (1616) or even that of Galileo Galilei (his abjuration in 1633 or his death in 1642). The different configurations make timescales that vary from 66 years to approximately 3 centuries.

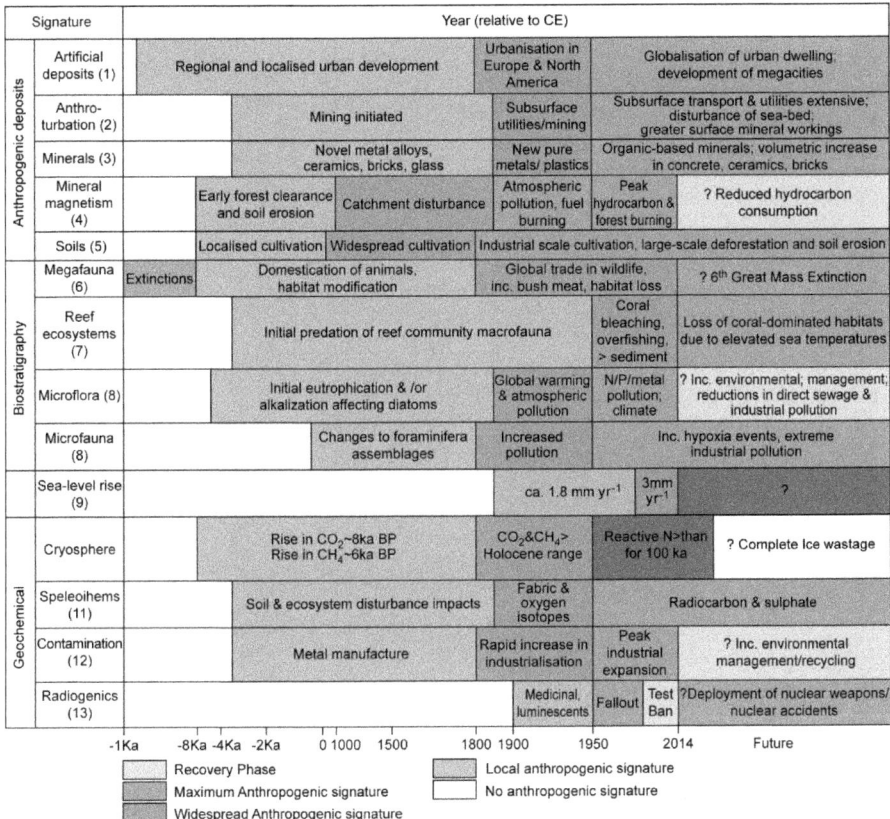

Figure 14.3. *Some anthropogenic signatures (modified according to Waters et al. [WAT 14]). For a color version of this figure, see www.iste.co.uk/grandcolas/biodiversity.zip*

Thus, Felisa Smith, from the University of New Mexico, Albuquerque [SMI 14, MAL 15], places the start of this epoch 14,000 years ago, at the colonization of North America by the first hunter-gatherers, as this colonization caused the disappearance of several large species of herbivore. These animals produced large quantities of methane that were released into the atmosphere, thus contributing to natural global warming; the diminution of atmospheric methane may have then led to the younger Dryas.

Several other types of information are used to characterize the Anthropocene: new materials (aluminum, concrete, etc.), new organic polymers, plastics, carbon microparticles and a whole array of chemical products, among which we find those linked to nuclear explosions (plutonium, C^{14}, etc.). The chosen criteria do not always coincide in terms of either time or place.

Thus, for example, if we take the impressions left by the extension of agriculture (Figures 14.2 and 14.3), they are manifest 11,000 years ago in Southwest Asia but only 2,000 years before our era in North Africa.

On the contrary, other criteria are global, such as the increase of the quantity of plutonium in sediments, but these modifications are so gradual that it is difficult to choose a limit that is not arbitrary (Figure 14.4). The same goes for modifications to the environment due to wooden fuel burned in order to melt metals during the Neolithic period, which differ according to the development of the populations in question.

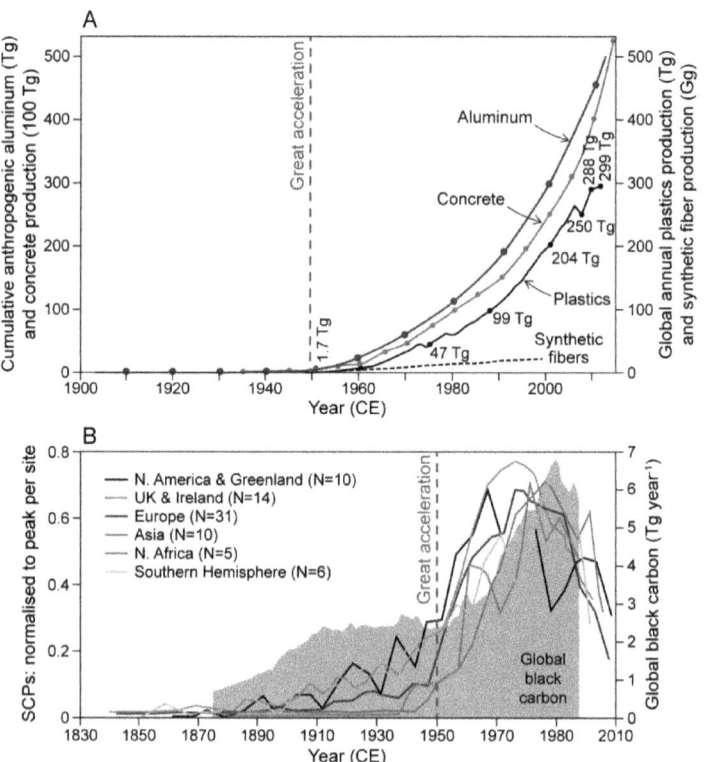

Figure 14.4. *Annual production of aluminum, concrete, plastics, synthetic fibers, carbon particulates and nitrogen fertilizers; all data that do not come from stratigraphy (according to Waters et al. [WAT 16], modified). For a color version of this figure, see www.iste.co.uk/grandcolas/biodiversity.zip*

The criteria associated with modifications in biodiversity vary in time: the decrease in vertebrate species has been observed since 1500, the decline in fish population since a century ago and the bleaching of the coral from 1979 (Figure 14.3). It would be paradoxical to gather under one category events that are not synchronous as markers in time!

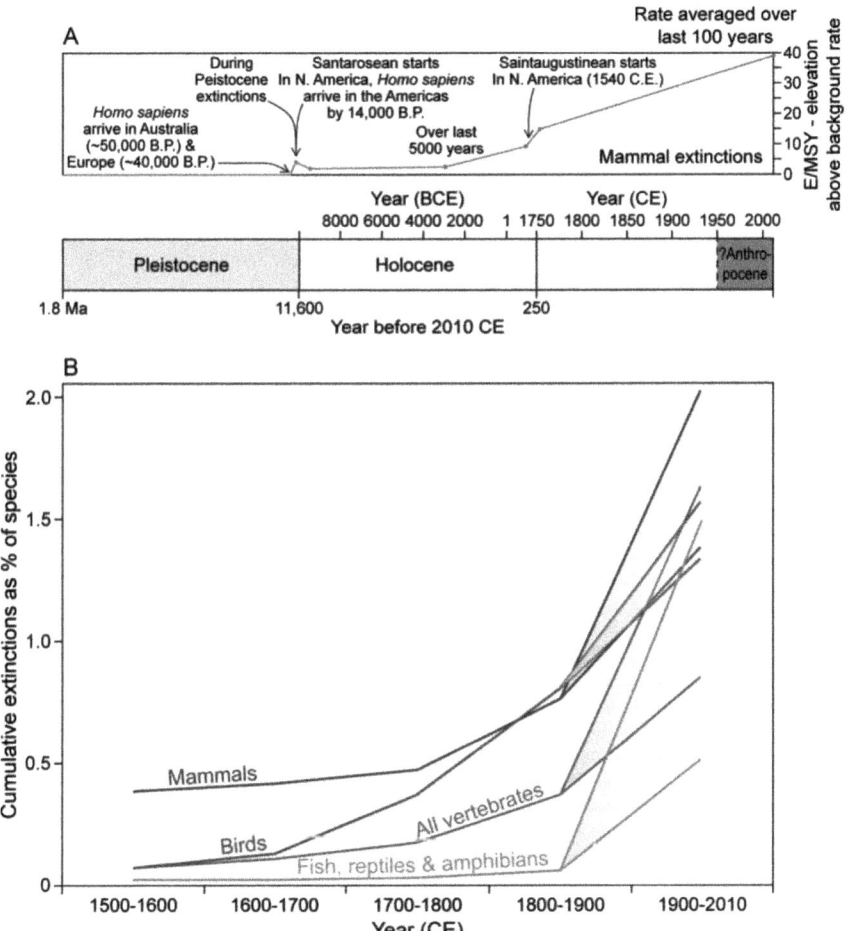

Figure 14.5. *Decrease in animal species. Data are mapped onto a Gregorian calendar. These data do not come from stratigraphic recordings by paleontologists; they are obtained rather by biologists, working from the basis of the loss of habitats of certain animal species (according to Waters et al. [WAT 16], modified). For a color version of this figure, see www.iste.co.uk/grandcolas/biodiversity.zip*

When aerial nuclear tests stopped, there was a drastic decrease in the quantity of plutonium (Figure 14.6). As this is a very marked decline, with the increase being more gradual, it was chosen by the task group assigned to the Anthropocene; however, this group has not, however, provided any stratigraphic data series, which is a prerequisite of the commission.

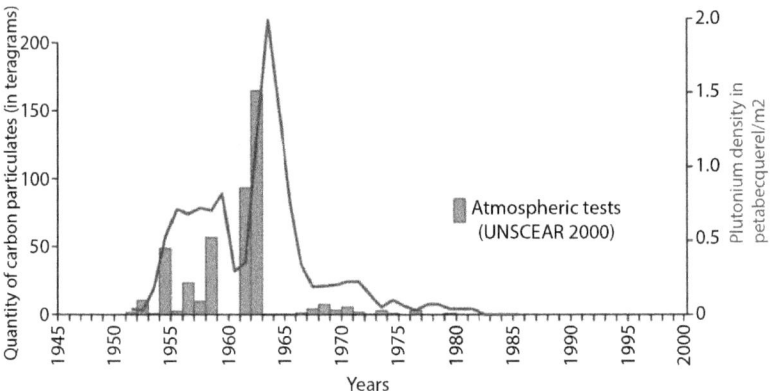

Figure 14.6. *The fall of the radiogenic signal as a marker of the Anthropocene. a) Quantity of C^{14} in the atmosphere before nuclear tests. b) Atmospheric concentration of C^{14} in green and plutonium (Pu$^{239-240}$) in blue, in relation to nuclear tests. PBq: petabecquerels (according to Waters et al. [WAT 16], modified). For a color version of this figure, see www.iste.co.uk/grandcolas/biodiversity.zip*

In fact, this data reference makes it possible to evaluate the potential correlation between different strata and, above all, between sediments deposited in different contexts (lake or marine environments, etc.). This comparison allows us to determine: (1) whether all these signals coincide and (2) their relative position compared with other sorts of indicators. It is disappointing that this point was not raised before the decision to choose this limit was made. Furthermore, there is no peak in 1945; in fact, the year chosen for the start of the curve for plutonium concentration in the atmosphere is 1950. This absence is troubling, as a point of reference that cannot be attributed to any single event cannot be considered as valid.

14.3. Why is the issue of the Anthropocene being imposed?

What is strange is that a group of scientists, not geologists for the most part, seem to want to force this subdivision onto the geological timescale. Recently, at the International Geological Congress held in Cape Town, from

28 August to 2 September 2016, only two speakers mentioned these stratigraphic phenomena. And yet, the session having barely started in the early afternoon, tabloids, at least in France, wanted to give a detailed account of this single aspect as if it was a key element of the Congress and the Anthropocene subdivision had been adopted.

The first speaker, Waters, defended the idea of introducing this period in the geological timescale (but, contrary to what has been said, he never proposed an era, only a "series").

The second speaker, co-author of this chapter, simply recalled the criteria required to introduce a new subdivision into the geological timescale and concluded that the Anthropocene did not meet these criteria. Finally, he remarked that no decision could be taken by the commission given that the task-group for the Anthropocene had not yet submitted a dossier to the sub-commission, which is the first step in the procedure!

This reason, therefore, entirely technical in nature, has nothing to do with the declamations of those who support the "geological force of the Anthropocene", such as Grinevald [GRI 12, p. 45], who claimed that geologists are more worried about "finding new raw materials, including energy sources, to support the development of industrialization, economic growth, etc". He also criticized the fact that this dossier is not on the agenda of the next International Geological Congress "despite the huge amount of pressure from the international press" (p.45) while just before he declared that "the dossier for the Anthropocene is far from complete, and not yet ready to be presented to the competent authorities at the International Union of Geological Sciences" (p.36).

14.4. Conclusion

Humankind exerts an influence on the planet's environment, which is not new; since the existence of life on Earth, it has modified the environment. It is life that led to the precipitation of huge quantities of iron at the bottom of the ocean, during the "Great Oxidation" 2.4 billion years ago. It is also life that, through its consumption of CO_2, brought the Earth into periods of total freezing, known as Ice Ages. What is perhaps new is that one species is responsible for the changes we see today, and the speed of change is greater.

It is also no novelty that humankind, which is building up its understanding, defining its terms, is attempting to place itself at the center of everything. We are a constant source of proposals and theories and thus always try to place ourselves at center-stage, which is not new. The intendant at Jardin du Roi, Georges Louis Leclerc, the Comte de Buffon, had already remarked in the 18th Century in *Les Époques de la Nature* (1778) that: "The entire face of the Earth now bears the mark of man's power". In the 19th Century, this idea was elaborated by the American ecologist George Perkins Marsh in his *Man and Nature, Physical Geography as Modified by Human Action* (1864) and, at the end of this century, the Italian geologist Antonio Stoppani suggested baptizing the most recent period "the Anthropozoic", in 1873. This term was also later used as an equivalent of the Quaternary.

Doubtlessly inspired by the French philosopher Henri Bergson, other more intellectualized approaches have attempted to highlight the novelty (supremacy?) of humankind by putting forward the term "noosphere" that now dominates the biosphere (Vladimir Vernadsky, Pierre Teilhard de Chardin, Edouard Le Roy).

Nobody can argue about the influence humankind exerts on the planet, but to make a new era of it, or just a new period, epoch or even stage, seems inappropriate. The timescale is not comparable in the slightest, 18 orders of magnitude separate them. What is a half-century or century compared to 26,000 centuries (which comprises a series, the shortest division)? So, is this a real phenomenon or an instance of human pretension? Is a lasting, planet-wide event sufficient reason to adopt a new era? If this is the case, it would be necessary to change era after large earthquakes as they change the distribution of masses and thus the position of the Earth's axis. For example, on 11 March 2011, an earthquake with a magnitude of 9^5 on the Richter scale occurred off the coast of Japan, followed shortly afterward by a powerful tsunami. This earthquake also modified the speed of rotation of our planet on its axis. A team of researchers from the Jet Propulsion Laboratory (JPL) from Pasadena in California calculated the repercussions of the earthquake on the angle at which the axis of the figure of the Earth inclines, its main axis of symmetry (around which terrestrial mass is balanced, not to

---

5 The power of an earthquake can be quantified by its *magnitude*, a notion first proposed by Charles Richter. Magnitude is not a scale but rather a logarithmic function. As a result, when the magnitude of the ground displacement is multiplied by 10, the magnitude increases by one unit: an earthquake of magnitude 7 will cause a ground displacement 10 times larger than a magnitude 6. The Richter scale is the most well-known term to laypeople.

be confused with its imagined axis of rotation from north to south). The particularity of earthquakes is that they slightly modify the distribution of our planet's mass according to their strength and where they occur. This phenomenon has already been observed following the terrible Sumatra earthquake in December 2004, which slightly modified the gravity of our planet. As a result, the day had shortened by 6.8 microseconds and the axis of the figure of the Earth had shifted by 8 cm. In the case of the earthquake in Chile on 27 February 2010, specialists estimated that the axis had shifted by 8 cm while the day had shortened by approximately 1.26 microseconds. Geophysicists predicted that the axis of the figure of the Earth had changed by approximately 17 cm and that the day had shortened by another 1.8 microseconds.

In the history of humanity, various different periods have been established, such as the Neolithic period and the Renaissance. The criteria used to establish these periods vary, as do their start and end dates, according to the chosen criteria. The subdivisions of the geological timescale are on the contrary based on a certain number of precise criteria. The ultimate goal is to accurately date. In the case of the Anthropocene, we know the dates of events to the year, sometimes to the day. There is therefore no use in introducing this period into the geological timescale, as has already been highlighted in several publications [KLE 15, FIN 16].

The period of the Anthropocene is defined as being caused by humankind, and it is written into the history of humanity, it has its place in the calendar of human history. Why, therefore, is there the desire to turn it into a geological era? This would be both futile and inappropriate, as it does not possess the characteristics of one.

14.5. Bibliography

[BER 95] BERGGREN W.A., HILGEN F.J., LANGEREIS C.G. *et al.*, "Late Neogene (Pliocene-Pleistocene) chronology: new perspectives in high resolution stratigraphy", *Geological Society of America*, vol. 107, pp. 1272–1287, 1995.

[BON 13] BONNEUIL C., FRESSOZ J.-B., *L'événement anthropocène*, Le Seuil, Paris, 2013.

[DES 29] DESNOYERS J., "Observations sur un ensemble de dépôts marins plus récents que les terrains tertiaires du bassin de la Seine constituant une formation géologique distincte; précédées d'un aperçu de la non-simultanéité des bassins tertiaires", *Annales scientifiques naturelles*, vol. 16, 171-214402-19, 1829.

[FIN 16] FINNEY S.C., EDWARDS L.E., "The "Anthropocene" epoch: scientific decision or political statement?", *GSA today*, vol. 23, nos 3–4, pp. 4–10, 2016.

[GIB 09] GIBBARD P.H., HEAD M.J., "IUGS ratification of the Quaternary system /period and the Pleistocene series/Epoch with a base at 2.58 MA", *Quaternaire*, vol. 20, pp. 411–412, 2009.

[GRI 12] GRINEVALD J., "Le concept d'Anthropocène et son contexte historique et scientifique", *L'anthropocène et ses issues*, Momentum institute Seminar, 11 May 2012.

[KLE 15] KLEIN G.D., "The "Anthropocene": what is its geological utility? (Answer: it has none!)", *Episodes*, vol. 38, no. 3, p. 218, 2015.

[MAL 15] MALHI Y., DOUGHTY C., GALETTI M. *et al.*, "Megafauna and ecosystem function: from the Pleistocene to the Anthropocene", *Proceedings of the National Academy of Science USA*, vol. 113/4, pp. 838–846, 2015.

[RUD 15] RUDDIMAN W.F., ELLIS E.C., KAPLAN J.O. *et al.*, "Defining the epoch we live in: Is a formally designated "Anthropocene" a good idea?", *Science*, vol. 348, no. 6230, pp. 38–39, 2015.

[SMI 14] SMITH F.A., "Recalibrating the Anthropocene: humans, megafauna and global biogeochemical cycles", http://oxfordmegafauna.weebly.com/ conference-blog/march-18th-2014 , 2014.

[SMI 15] SMITH F.A., ELLIOTT S.M., LYONS S.K. *et al.*, "The importance of considering animal body mass in IPCC greenhouse inventories and the underappreciated role of wild herbivores", *Global Change Biology*, vol. 21, pp. 3880–3888, 2015.

[WAT 14] WATERS C.N. *et al.* (eds), *A stratigraphical basis for the Anthropocene?*, Geological Society, Special Publications, vol. 395, London, 2014.

[WAT 16] WATERS C.N. *et al.*, "The Anthropocene is functionally and stratigraphically distinct from the Holocene", *Science*, vol. 351, no. 6269, 2016.

List of Authors

Guillaume ACHAZ
ISYEB, UMR 7205 CNRS
UPMC EPHE
National Museum of Natural
History
Sorbonne University
Paris
France

Eric BAPTESTE
ISYEB, UMR 7205 CNRS
UPMC EPHE
National Museum of Natural
History
Sorbonne University
Paris
France

Serge BERTHIER
Paris Diderot University
Institut des Nanosciences de Paris
(INSP – UPMC)
Cape Town
South Africa

Patrick BLANDIN
ISYEB, UMR 7205 CNRS
UPMC EPHE
National Museum of Natural
History
Sorbonne University
Paris
France

Nicolas CHAZOT
Department of Biology
University of Lund
Sweden

Claudine COHEN
Ecole des Hautes Etudes en
Sciences Sociales
Paris
France

Patrick DE WEVER
CR2P, UMR 7207 CNRS
UPMC EPHE
National Museum of Natural
History
Sorbonne University
Paris
France

Vincent DEBAT
ISYEB, UMR 7205 CNRS
UPMC EPHE
National Museum of Natural
History
Sorbonne University
Paris
France

Sébastien DUTERTRE
Institut des Biomolécules Max
Mousseron
UMR5247 – CNRS
University of Montpellier
France

Marianne ELIAS
ISYEB, UMR 7205 CNRS
UPMC EPHE
National Museum of Natural
History
Sorbonne University
Paris
France

Stan FINNEY
California State University
Long Beach
USA

Philippe GRANDCOLAS
ISYEB, UMR 7205 CNRS
UPMC EPHE
National Museum of Natural
History
Sorbonne University
Paris
France

Doris GOMEZ
Centre d'Écologie Fonctionnelle
et Évolutive UMR CNRS 5175
Montpellier
France

Thomas HEAMS
UMR GABI, AgroParistech
INRA
University of Paris-Saclay
Jouy-en-Josas
France

Jean-Pierre HUGOT
ISYEB, UMR 7205 CNRS
UPMC EPHE
National Museum of Natural
History
Sorbonne University
Paris
France

Aude LALIS
ISYEB, UMR 7205 CNRS
UPMC EPHE
National Museum of Natural
History
Sorbonne University
Paris
France

Jean LEHMANN
Institute for Integrative Biology
of the Cell (I2BC), CEA, CNRS
University of Paris-Saclay
Gif-sur-Yvette
France

Violaine LLAURENS
ISYEB, UMR 7205 CNRS
UPMC EPHE
National Museum of Natural
History
Sorbonne University
Paris
France

Philippe LOPEZ
ISYEB, UMR 7205 CNRS
UPMC EPHE
National Museum of Natural
History
Sorbonne University
Paris
France

Jessica MARTINEZ-VARGAS
Departament de Biologia Animal,
de Biologia Vegetal i d'Ecologia
Facultat de Biociencies
Universitat Autonoma de
Barcelona
Spain

Marie-Christine MAUREL
Institut de Systématique,
Evolution, Biodiversité
ISYEB-UMR 7205-CNRS
MNHN, UPMC, EPHE
Sorbonne Universities
Paris
France

Romain NATTIER
ISYEB, UMR 7205 CNRS
UPMC EPHE
National Museum of Natural
History
Sorbonne University
Paris
France

Violaine NICOLAS
ISYEB, UMR 7205 CNRS
UPMC EPHE
National Museum of Natural
History
Sorbonne University
Paris
France

Nicolas PUILLANDRE
ISYEB, UMR 7205 CNRS
UPMC EPHE
National Museum of Natural
History
Sorbonne University
Paris
France

Carole SAINTOMÉ
UMR 7196 CNRS, U1154
INSERM
National Museum of Natural
History
Sorbonne University
Paris
France

Chloé VIGLIOTTI
ISYEB, UMR 7205 CNRS
UPMC EPHE
National Museum of Natural
History
Sorbonne University
Paris
France

Thierry WIRTH
ISYEB, UMR 7205 CNRS
UPMC EPHE
National Museum of Natural
History
EPHE – PSL University
Sorbonne University
Paris
France

Index

A, B, C

amino acid, 89
archetype, 6–9, 11
Bayesian, 46, 128, 129, 133, 204,
  205, 217
bioengineering, 17–21, 26
biogeography, 180
biotechnology, 17, 26
birth–death, 40
butterflies, 139–141, 147, 150–152,
  156–161, 163
Charles Darwin, 1
climatic, 221
coalescent, 40, 46, 48–51, 53–55
codon degeneracy, 96
color pattern, 150, 153, 156, 158
comparative biology, 29
cone snail, 227, 230, 234–236,
  239, 241
conotoxins, 232–234, 238–243
convenience science, 64, 65, 68, 69,
  71, 79
CPR bacteria, 67
Creationism, 1, 2, 11
criteria, 253–255, 258, 259, 261, 263

D, E, F

dating, 203
demographic, 203–205

diversification, 149, 156
drug-resistant, 124
ecology, 23–25, 140, 150, 152, 155,
  160, 163, 164
ecosystem, 15, 16, 18, 23–26
embryology, 3, 4, 9–11
ethics, 33–36
Europe, 213
evolutionary
  biology, 39
  process, 41, 53, 55, 56, 177, 178, 180
  theory, 8
evolvability, 18
extinction, 43, 44, 46, 56
  crisis, 33, 36
flight, 152, 156–159, 162–164
France, 215–218, 220, 221

G, H, I

G-quadruplex, 112, 113, 120
genetic
  data, 214, 218–220, 223
  drift, 90, 101, 104
  system, 94, 100
geological scale, 256
host, 190, 192–197, 199, 201, 202, 206
humankind, 254, 255, 261–263
integrated approach, 205

J, K, L

Johann Wolfgang von Goethe, 4
kinship, 40, 41, 46, 49, 50, 53–55
Lassa virus, 189
lineage, 41–46, 48, 49, 51, 54, 56
linkage bloc, 51, 52
living world–machine, 17, 18, 20, 21

M, N, O

markers, 259, 260
metagenomics, 63
microbiome, 63
microbiota, 64–67, 70, 72, 73, 75, 79, 80
mole, 213
molecular, 194, 203, 204
morphology, 3–11, 214–216, 218, 219, 223
Mycobacterium tuberculosis, 123, 125
nanostructure, 157
natural theology, 2
Naturphilosophie, 3–6
niche, 221–223
OB-fold proteins, 115
origin of biodiversity, 1
OTU, 69, 72, 74

P, Q, R

path dependency, 68, 69, 71, 72, 75
phenotypic plasticity, 200
photonics, 147, 148, 157
phylogenetic, 29–32, 34, 35
    diversity, 31, 35
    trees, 31, 32, 35
phylogeny, 177, 178, 180, 227–230, 237, 238, 242, 243
protein structure, 102
proteome, 89
proteomic, 238, 239
Quaternary, 252, 253, 262

rare biosphere, 67
rational design, 16, 18, 20, 25
reverse transcriptase, 116, 117
ribonucleoprotein complex, 119, 120
ribosome, 93–98, 100, 102, 104
Richard Owen, 1

S, T, V, W

selection
    natural, 155, 157, 160–162
    sexual, 160–162, 164
shelterins proteins, 114, 115
Spain, 215, 216, 218, 221
speciation, 42, 43, 46
species, 30–35
    delimitation, 178
    distribution, 221, 222
spread, 124, 127, 131, 133
standardization, 18
Staphylococcus aureus, 123, 130
stochastic, 41, 43, 53
stratigraphy, 251, 252, 254, 255, 258–261
subdivision, 251, 252, 254, 255, 260, 261, 263
synthetic biology, 19–22, 24, 25
systematics, 31
t-loop, 113, 114, 120
taxonomy, 176, 178, 180, 182, 219, 223
telomerase, 109
    biogenesis, 119–121
telomere, 109
TER RNA, 116, 117, 119
timescale, 251–256, 260–263
transcriptome, 238, 240
translation, 90, 91, 94, 98–100
tRNA, 95, 96, 100
turrids, 227–231, 237
typhoid fever, 132
venom, 227–234, 237–243
wings, 139–142, 144, 147–157, 159, 163

CPI Antony Rowe

Chippenham, UK

2018-03-27 09:38